Physics and Basketball

The Student Athlete's Guide to Improving On and Off the Court

By Dorian Bohler

COPYRIGHT

All rights reserved.

TABLE OF CONTENTS

Acknowledgements

In 2009, I wanted to write a book to help students who lacked access to adequate schools and resources obtain a world-class education. At that time in my life I had two years of experience as a teaching assistant at the university level and had tutored privately for many years.

I discussed this project with my good friend who has been a public school teacher for nearly 20 years. He urged me to set the project to aside and gain some experience working directly with the students that I wanted to impact the most. That year I began working as a mentor and it was one of the most rewarding decisions I have ever made.

At the time of writing this text in 2018, I've had so many great experiences working with students in New Jersey, Michigan, Georgia, and California. I've seen many of them matriculate from junior high all the way through college. What is most special is that these boys and their parents have allowed me to play a role in their successes, achievements, and ambitions. These experiences have made me a better man and have laid the foundation for this work.

It has also been a great accomplishment to receive so many awards for my work in the community. I'd like to thank Kappa Alpha Psi Fraternity, Inc. and my fellow members across the world who support our many youth programs. The training, support, and encouragement ultimately led to the success of our youth leadership program for young men of color in Oakland, California, and surrounding areas.

Lastly, I'd like to thank the many members of my family who continue to share so many strong words of encouragement. I'm grateful to have a mother who has always urged me to give more of myself. My two sons, Ethan and "DJ" have inspired me to be more creative and patient. My wife, Erica, continues to reassure me that the time I spend away from the family is a worthwhile investment. She is also kind enough to use much of our time together to help plan each speech, student lesson, or major initiative. To have so many supporters is truly a blessing.

Chapter 1: Why Am I Writing This Book?

Why am I writing this book? Primarily, it's because I have spent the past 10 years working with students just like you. I'll take my story back to my days as a graduate teaching assistant at Wayne State University where I noticed that even though so many "capable" students make it to college, few were truly enthusiastic about learning—especially as it related to their physics coursework.

It was abundantly clear to me that students who were highly intelligent, some with goals of going to medical school, really didn't connect with the information. Fundamentally, their main problem was that they were taking these science classes just for the grades so that they could move on. Ironically, this seemingly pragmatic approach rarely served them well, and the students would often do poorly in my class.

More recently (for the past eight years or so), I've been working with various nonprofits as an advocate for science education and leadership skill development. In 2013, I started a youth leadership program that trains young men how to succeed in life and sidestep many of the pitfalls that have led so many others down the wrong path. I mostly enjoy working with young people who have a lot of potential but not many resources to help them realize it. As a young kid from Detroit, there weren't a lot of people willing to help me but I wouldn't be where I am now without the help of others, so I try to pay it forward.

I'd say less than 10 percent of the students passing through my program have any exposure to physics before they take the subject in high school—which generally isn't until the 11th or 12th grade. I think this is really too late, so I've made it my mission to turn this around for young people.

Most people don't meet many physicists, so I'm constantly invited speak to audiences mostly consisting of high school and college students. But over the course of my conversations with young people, one thing that really struck me is that they are so passionate about sports. Seems like a no-brainer, right? But a bit discouraging for a science evangelist.

Riffing on this, I thought back to my own youth and some of the things that inspired me. Like those kids, one of my favorite things to do was to play sports, particularly basketball. In fact, I still play a few times a week and it's really one of my favorite things to do in life. So I tried a little experiment with some of my students, using basketball to get them more excited about learning. Bingo! Student engagement and enthusiasm shot up. It should have been obvious because that's what hooked me as a kid. Linking the things that I enjoyed—like basketball—to engineering programs and other extracurricular activities made it so I couldn't wait to hit the courts after doing something productive on Saturdays. So that's why I'm writing this book—I want to give you the opportunity to take what you love to do and learn something new at the same time.

Why physics, you ask? Physics is a great subject for students to be exposed to. I can speak to this personally, as it has been very rewarding for me from a career perspective. As a scientist, I have worked at two of the top universities in the country, Stanford and Princeton, and have had the opportunity to travel all over the world. But even here there's a secret connection to basketball. When you travel, you meet different kinds of people, and one thing that never fails to tickle me is that there's a local pickup game at each one of these places. No matter where I am, one of the best ways to connect with new people is through my love of the game. There are scientists from Germany, China, the Middle East, Europe, and South America, who love shooting hoops as much as they can so that allows us to create bonds outside of the office, which helps us work better together.

Of course, as with many things in life, basketball is more fun once you learn the basics—if you can dribble with both hands, and you have some decent skills, then you can really enjoy the game a lot more. It's the same with physics, and science in general. If you can grasp the fundamentals to inform your thinking, you can start to dig into the nuances of life and tap into your potential; however, a lot of people never overcome that barrier because they can't get over the first hump. Maybe they have no interest in the subject because of how it's presented, or they have negative experiences regarding education and walk into the classroom in the wrong state of mind.

When I first started training and mentoring students long ago I had no clue that I'd discovered what would become my life's passion. It's so rewarding to help young people who ultimately end up graduating from college and pursuing their dreams. However, I must admit that it can really get me down when I see good people achieving far below their potential. In this vein, I recently watched a Netflix docu-series called *Last Chance U*. The show depicts college football players who have had to leave, or never had the chance to attend, Division 1 schools, and are instead playing football at a junior college in Alabama. This team is their last chance to make an impact on football scouts, to have a shot at playing D1 so they can go on to play on professional teams.

And this show, as entertaining as it was, hurt me to the core. I was disappointed that the show portrayed the students, most of whom are black, as deeply ignorant. *Last Chance U* unfortunately reinforces some long-standing stereotypes. And what you end up seeing through that show is the worst-case scenario when it comes to education in the United States, particularly for education and sports packaged together for blacks in this country.

What was so disappointing was that the students depicted as "last-chance kids" had such a small view of the world. In particular, they *felt* as if their only opportunity for success in life was dependent upon a rubber ball. So much so that many of them completely blew off the entire process of education or anything outside their bubble. It's a subject of another book to investigate how this poisonous mindset takes root. The series really inspired me to double down on my efforts to use sports to prevent students from going down the same path. I want them to realize that regardless of their background, race, or income level, and what people say or think, that the potential to learn and do great things is in them. I want to encourage students across the world to use their own natural curiosity to unlock their potential.

A tragedy that I often see in student-athletes, or students in general, especially those lucky enough to go to college, is that they don't take full advantage of their opportunities. That is really sad because at that stage in your life the world should be your oyster and you should be arming yourself with everything that you need to change your trajectory for the better!

Challenges that Students Face

In my experience observing students, the lack of curiosity leads to a lower quality life because they adopt a philosophy which is comfortable with not grasping a myriad of subjects which are comprehendible. That lack of curiosity steals from them in so many different areas: it limits the size of their social circles, diminishes opportunities to do business, reduces the amount of life experiences which may be interesting, and so on. Curiosity is like a muscle, if it's not used and strengthened, then it atrophies and dies. Science is a particularly good exercise for curiosity because it gives an individual a very tangible way to use their senses and strengthen that "muscle."

In addition, I see too many students affected negatively by lack of character. When we talk about sports, most athletes realize the value of good leadership and yet it's not very common. Beyond that, there are so many layers to having a good character such as being honorable, disciplined, and hard-working, which are essentially all the things your grandmother knew were important.

Yet another place where students often face challenges, and boy does this resonate in my own life, is that lot of schools just *suck*. Finding a good teacher is very hard, and a lot of people just don't have the option of going to decent schools. Obviously, there's a correlation between wealth and school quality, but what's so interesting about today's world is that the Internet gives us all access to the same basic information. So one might argue that in this day and age, a student who is deficient in a subject doesn't have a good excuse for not overcoming his learning barriers because the Internet makes it about personal accountability. I know there are many more factors to consider when we start talking about self-teaching, the Internet, and the education system. However, in the rest of this book, I will share tools that are available to better yourself, and also methods that you can use to approach learning if you are faced with some difficulty in grasping a subject. Fundamentally, it amounts to personal accountability, goal setting, and a willingness to go the extra mile in all phases of life.

When it comes to academic performance, a lot of students never really learned *how to learn.* Or, to put it differently, many of them are unaware of how to improve in subjects that require skills that lie outside of their current levels of ability. Student-athletes are special because they are adept at overcoming physical challenges. So now they have to take what they know about improvement on the court and apply that same grit to being successful in the classroom.

Another issue student-athletes face is having a poor (or nonexistent) life plan. I've never been one to say, "Hey, you want to go to the NBA? Do you want to go to the NFL? You're not going to make it anyway, so you might as well study physics." That's not really my point. My point is that when an individual approaches their life and chooses their goals, they should have a risk-management approach in which they formulate what they're going to do. So, when you say you want to be a sports professional or if you want go to college and play at a collegiate level, I'm not necessarily saying you need a backup plan but you do need a series of skills that will give you options later. And

I can't think of a better skill to have than being comfortable with technology and mathematics! And guess what? The study of physics is foundational for understanding technology and mathematics.

Sports as Business

I'd like the "last-chance students" who love playing sports, and think their only opportunity at the good life is playing professional ball, to take a wider view of the sports industry. There are unlimited opportunities to earn income in the sports industry and capitalize on new breakthroughs in technology.

Many athletes from low-income backgrounds are working really hard at their sport while trying to overcome so many social barriers like equal access to education, employment, health care, and so on. They too should take a look at their sport from top to bottom. In addition to being competitive on the field, more athletes are beginning to focus on self care and health. Some others are taking more risks in business by galvanizing their own branding. There are also technology opportunities related to manufacturing the equipment that is used. Moreover, as we continue in this new era of technology revolutionizing performance and training enhancements, having a good grasp of physics will improve your ability to grasp a lot of these tough concepts in the future.

Growth

After all, life is a series of situations that require you to grow. Until the day we die, we're going to meet many situations for which we are simply not good enough and we must have that growth mindset. But to be successful, you can't just have the mindset, you need a growth *skill set* of knowing how to approach the problem and how to structure your life so that you can learn.

Physics class is a great place to pick up this growth skill set because it is a very tough subject. It's not really about the problems themselves, it's more about who you have to become in order to gain the ability to get the right answers. You have to develop mental discipline, you have to think creatively with great precision, and you have to develop curiosity about the smallest details. These are all qualities that are important for a person who wants to live a successful life.

Recently, student-athletes have been graduating from college at much higher rates. However, I read a study that detailed six-year college graduation rates segmented by race. Unfortunately, Blacks and Hispanics graduate at a rate of 37% and 46%, respectively, 20% below their White and Asian counterparts. As I write this, I find myself wondering how many of those dropouts were athletes. In addition, I know the chances of finding well-paying, fulfilling work without a college degree are greatly diminished.

One recent success story involves a current professional basketball player for the NBA champion Golden State Warriors. Before he was given the nickname "Swaggy P," Nick Young grew up in a pretty tough environment. His family dealt with the loss of his older brother to gang violence, and that turned his world upside down at an early age. Despite his struggles, he averaged 27 points per game as a senior in high school and was highly recruited. One poll had him ranked as the seventh-best player in the country. In order to move on to college, Young had to do well enough on standardized tests. Unfortunately, this was a huge challenge for Young as he had been placed in special education. Fortunately, he had the support of his entire school—his counselors, his tutors, and so on—so he buckled down and finally did well enough to play at the next level.

Though Young's story is certainly inspiring, my question is, "What happens to the guys who didn't pass, like those guys who end up in *Last Chance U*, who didn't have the grades?" What can we do to make sure that more student-athletes grasp the importance of academics?

The statistics say that students who take quantitative advanced placement (AP) courses like physics, calculus, or statistics do much better on standardized tests and are accepted into better colleges. I would like to set that as a goal for my readers so that student-athletes aren't only planning to do well in sports, but are striving for excellence in academics.

Other Positive Effects

Studies show that activities such as learning mathematics, chess, or how to play an instrument, all affect how your brain actually works. The neurons in your brain form new neural networks and improve your overall brain efficiency, which might help you in your sport. Of course, this brain development will benefit you in many other areas of your life. So the take-home point here is that even if you have little interest in a physics class, buckle down, work hard, and expand your horizons. Doing so will slowly change the structure of your brain, and with that comes many other benefits.

Too many people have trouble starting and completing challenging tasks. Even though, they have plenty of evidence that these undertakings are possible. A serious study of physics gives you the opportunity to approach a wide variety of problems and see them all through to the end. As you solve problems in physics, you begin to develop a systematic way of thinking through challenges. You realize that finding answers is not just routine "plug and chug" but requires thought, perhaps even research, and certainly collaboration with your classmates. So you approach a problem, you overcome the challenge, and you see it all the way to completion! Even then, you think about what you could have done to get to that solution a bit quicker. What a beautiful analogy for personal development for life. So much can be learned from this process.

Your horizons really expand once you get into the details of physics and begin to understand some of the subdomains such as mechanics or astronomy. It's truly *fascinating* to consider how small we really are and the various phenomena at work in the universe. If you could only get past trying to get the answer for the sake of getting the answer, and instead took the opportunity to be fascinated by the ways that physicists in the past have actually arrived at these great breakthroughs, that would be so beneficial to you!

Charlie Munger, a billionaire and business partner of Warren Buffett, has a very interesting concept he calls "the toolkit." The toolkit analogy pretty much goes like this: to a person with a hammer, everything looks like a nail. So you're trying to solve all these different problems in life, and if all you have are sports, then your options are surely limited. Picture the stereotypical father with a limited outlook on life, who uses sports as an analogy to explain every single thing related to his child's life. It's a punchline for a reason! If we then can expand our toolkit and put a little physics in there, as well as some statistics, then a bit of computing, followed by some economics, we would then be empowered to approach so many of life's challenges with a different perspective that might allow us to do some beautiful things, and make some strong contributions to the world and humanity.

Ultimately this book is about success. It will show students the benefits of learning physics and other tough subjects. I use physics because it's the avenue I used to widen my view of the reality. You will learn a systematic approach to improvement, which will allow you to succeed in any field of your choosing!

Chapter 2: Physics Primer

Let's begin our study of physics by discussing mechanics! You may find yourself confronted with a lot of new words and concepts in this chapter and the next one. Get out a pen and paper and write down concepts you're unfamiliar with, then go learn more about them from a textbook, a good teacher, Google, or YouTube. Don't be intimidated—you've got this.

But back to mechanics. This foundational branch of physics describes the motion of objects and the forces acting on them. In the nice, simple world of mechanics, you are usually given a system with one or two objects, and asked to figure out how the object moves with time. The mathematically precise description of how the object(s) move with time is called an **equation of motion** *(EOM)*. To solve many mechanics problems, the main goal is to find the system's EOM. Let's work out an example. Take a function of the form:

$$x(t) = v_i t + 4$$

where x is the position (in feet) of a ball at various time points (measured in seconds) *t*. Here, the EOM was given to us already. If we're given the velocity v_i we can solve the ball's position for all times *t*. Say at $t = 0$ the ball was initially at position $x = 4$, 20 seconds later (at $t = 20s$). Moving at speed of $v_i = 3\frac{ft}{s}$ the position becomes

$$x(20) = 3 * 20 + 4 = 64 ft.$$

But what do we do when the equation of motion wasn't given to us? For this, we need to understand *forces.*

Forces

Very simply stated, *a force is a push or a pull by one object on another.* Boom. Done? Not quite. In physics, we examine situations a little more precisely, so let's spend some time understanding forces. Even without any formal training in physics I'm sure that you have heard of many types of forces such as weight and friction. Terms such as drag and the normal force might be less familiar to you.

Before getting ahead of ourselves, let's point out that motion of an object is always the direct result of a force acting on it. That basketball traced a beautiful arc as you made a 3-pointer because you *pushed* it, and gravity *pulled* the ball back toward the earth. A 17th-century scientist named Isaac Newton gave us a way to precisely formulate this intuitive reality. **Newton's first law of motion** states that in absence of an external force an object at rest remains at rest, and an object in motion continues in motion with a constant velocity.

The tendency of an object to resist any attempt to change its velocity is called **inertia**. A key concept related to Newton's first law (and basketball) is the **inertial reference frame.** A reference frame is simply a system of coordinates that describe position points relative to a body. An example of this is degrees of longitude and latitude measured from the equator and the earth's poles, which describe points of positions on the earth's surface. An inertial frame of reference is simply one which isn't accelerating. According to the first law, bodies at rest and constant velocity are in equilibrium. A passenger in a car traveling 50 mph can pour a cup of coffee as easily as a person sitting stationary in their kitchen. However, if the driver steps on the gas, the car accelerates. This makes the reference frame *noninertial*, therefore the first law no longer applies and the coffee will spill directly onto the passenger's lap.

Newton's second law of motion states that a force (remember, *push or pull*) acting on an object is equal to the mass of the object m multiplied by the **acceleration** of that object a. Acceleration is the rate of change of velocity. Think of slamming on your brakes. The velocity of your car changes very quickly, right? That's a lot of (negative) acceleration and you and your frightened passengers certainly feel a force! That scenario is Newton's second law in action! The units of acceleration are $\frac{ft}{s^2}$ and seem like nonsense until you remember that **velocity** is the rate of change in position with time (how fast an object is moving) and the velocity has units of $\frac{ft}{s}$. In fact, as long as we maintain the same dimensions, namely distance and time, we can use any units, such as meters per second or miles per hour.

So naturally, the rate of change of velocity (how quickly an object's velocity is changing) will have units of feet per squared or $\frac{ft}{s^2}$.

All of this talk of rates might have triggered your calculus brain. Physicists often use calculus, (the mathematical study of continuous change) to compute the position from the acceleration. In fact, the mathematical language of calculus was actually developed to study nature, so you could say that calculus was developed specifically for physics. Since calculus can be a bit tricky, many introductory physics courses often make simplifications and eliminate the need to do calculus altogether.

Gravitational forces exist between any two massive objects (no yo' mamma jokes, please). The magnitude of the gravitational force is given by the mass of object, written m_1 and m_2 in the following formula:

$$F = G\frac{m_1 m_2}{r^2} \quad [2.1]$$

where G is the gravitational constant 6.67×10^6 $m^3/kg/s^2$. Let's work out an example close to home: earth has a mass of M = 5.972×10^{24} kilograms and a radius of 6.367×10^6 meters. The force on an object of mass m is given by:

$$F_g = G\frac{Mm}{r^2} = gm = W \quad [2.2]$$

where g is the gravitational acceleration on earth, 9.8 m/s². When you multiply g by m, you obtain W, the object's weight.

In physics the word "normal" means perpendicular, or 90 degrees to a surface. So, a **normal force** arises every time two surfaces are in contact. If a basketball is on the floor a normal force pushes up from the ground to counteract the gravitational force.

The **force of friction** counteracts motion in a particular direction, and becomes important when two surfaces are in contact. The force of friction opposes the motion such that it causes motion to slow down. Think of pushing a heavy dresser against carpet; that extra pushing you have to do is to counteract the force of friction. The pushing you have to do up until the dresser just starts moving is called **static friction**. After the dresser starts moving, but you still have to keep pushing that stupid dresser, (harder than you would on, say, a hardwood floor), **kinetic friction** has taken over. Both static and kinetic friction forces are proportional to the normal force between surfaces (dresser and carpet) and can be related by their respective *friction coefficients* μ_k and μ_s. $F_{max} = \mu_s N$ and $F = \mu_s N$. F_{max} describes the situation where if the force is greater than F_{max}, the object (dresser) will start to **slip** across the carpet.

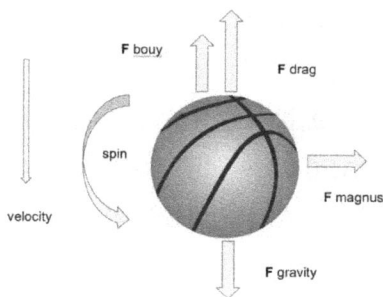

Figure 2.1 Types of forces on a basketball which is falling and spinning

If we look at the forces on a basketball in free fall which is also spinning as shown in Figure 2.1, you will note that there are a few more forces that we haven't mentioned.

The **buoyant force** on the ball is the total upward force caused by collisions of air molecules with the outside surface. This is the same force that allows boats to float on water. **Drag force** acts on the ball when moving and is essentially friction caused by air acting in the opposite direction. **Magnus force** arises because the ball is moving and spinning with essentially the same force as the force that makes a baseball curve. This force occurs because of the variance in drag on different sides of the ball due to its spinning motion.

Ok, now you know something about all the forces so how do you actually solve your homework? Typically, a great starting point for solving mechanics problems is to draw a free body diagram which has labels that indicate all the forces vectors acting on a particular object as well as a coordinate system which shows the positive x and y directions. This also includes summing all the forces acting in each direction.

Vectors

Forces are **vector** quantities, which means that they have both a number to quantify their size and a particular direction. The size of a vector is called its **magnitude**. We refer to numbers which only have magnitudes, not directions, as **scalars**. You can think of scalars as regular numbers since they describe things like distance, mass, and temperature (where the heck does 85^0F point?). However, a vector has a direction—"the suspect is headed east at 97 miles per hour." Jokes aside, the main takeaway here is that forces and accelerations must have a direction! For example, if two forces have the same magnitude but act in opposite directions, the sum of the two forces is zero.

However, if they act in the same direction, the net force on the object is doubled. In the former case the object would be stationary, but in the latter case the object may move. We denote the fact that vector quantities, like forces, have direction by writing little arrows over the top: $\vec{F}, \vec{a}, \vec{v}$ for the force, acceleration, and velocity vectors, respectively. Sometimes textbooks will also write vector quantities in bold.

Vectors can also be broken down into their components in each direction. Figure 2.2 shows a diagram of a velocity vector resolved into its vertical and horizontal components.

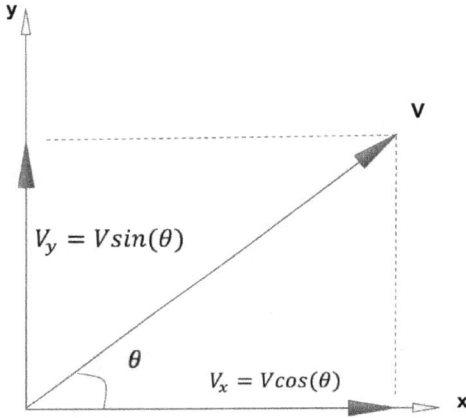

Figure 2.2 Vector in component form

Linear Motion

Now that we know that forces have direction, how do we figure out the motion of an object in our system? How do I compute my basketball's trajectory? To find EOM and do your physics homework, begin with accelerations. A slightly different version of Newton's second law,

$$\sum \vec{F} = m\vec{a} \quad [2.3]$$

may be used to solve for an unknown acceleration with the form

$$\vec{a} = \frac{\sum \vec{F}}{m} \quad [2.4]$$

which could then be used to find the EOM. If you aren't familiar with Σ, it is mathematical shorthand for "the sum of" or "the sum over." So, if you have a few forces acting on an object, ΣF just means "take the sum of all those forces."

To get an EOM, you frequently need to determine $v(t)$, the velocity of the object at a later point in time. If we know the object's initial velocity v_i then we may use what is known as **equations of kinematics** to determine $v(t)$. Kinematics means "the geometry of motion," and its well-known equations allow us to solve many problems without needing to deal with masses and forces. As we discussed before, acceleration is the change in velocity. If I know both v_i and the acceleration a, then the velocity at all future times $v(t)$ is given as:

$$v(t) = v_i + \int_0^t (a)dt \quad [2.5]$$

The swoopy integral sign $\int$ introduced above simply means the *continuous sum* or total quantity over a given range. In our initial example this could be from $t = 0$ to $t = 20s$. By the way, you've just seen a glimpse of calculus. Congrats if this is your first time!

Let's put acceleration and velocity in terms of an example from basketball. Consider the function $p(t)$ to describe the number of points Steph Curry had scored at any particular point in time during his basketball career, and let the function $q(t)$ describe the number of points scored by Curry in an individual game.

The function $q(t)$ describes the change in the function $p(t)$ in precisely the same way $a(t)$ describes $v(t)$. Knowing the number of points Curry has scored before the 2018 season, we can calculate the balance at the end of the season-opening game as follows:

$$p(t) = p(0) + \int_0^t q(t)dt, \qquad [2.6]$$

where t_f would be 48 minutes, the number of minutes in a basketball game.

Going back to our previous velocity example, we could repeat the integration one more time to find the *position*. In other words, we find the position at time t by summing up all the rates of change of position (which is the velocity):

$$x(t) = x_i + \int_0^t (v)dt \quad [2.7]$$

So, we may predict the position of the objects by remembering the following relation:

$$\frac{1}{m}\sum F = a(t) \rightarrow v_i + \int dt \rightarrow x_i + \int dt \rightarrow x(t) \quad [2.8]$$

This relation sums up everything we've discussed so far; it's not new physics. We can find the acceleration by summing all of the forces on a particular object and dividing by the mass (Newton's second law). We can then find the equation of motion by taking two integrals, one over acceleration, one over velocity, as long as we know the initial velocity v_i and position x_i of the object.

Studying mechanics is essentially understanding how to identify which forces contribute to a particular system, as well as understanding how those forces add together in order to cause acceleration. Even if an instructor gives us a tricky problem, they must generally give us enough detail to determine initial parameters x_i and v_i, which means we can solve for the EOM. Below you'll find three equations describing an object undergoing constant acceleration.

$$a(t) = a \quad [2.9]$$
$$v(t) = at + v_i \quad [2.10]$$
$$x(t) = \frac{1}{2}at^2 + v_i t + x_i \quad [2.10]$$

By combining equations, we can also obtain the relation for a final velocity based on an initial one, and final and initial positions.

$$v_f^2 = v_i^2 + 2a(x_f - x_i) \quad [2.11]$$

We will refer back to these equations in the rest of the chapter and also other chapters in the book.

Projectile Motion

Now things are getting interesting, as we move from one spatial dimension to two. Here, we'll use position vectors, $x(t)$ and $y(t)$, to denote horizontal and vertical positions, and velocity, $v_x(t)$ and $v_y(t)$, and acceleration, $a_x(t)$ and $a_y(t)$.

An object is said to be in **free fall** if the only force acting on it (while it's falling) is the force of gravity, as in equation 2.2. On earth, the force of gravity produces a constant acceleration $g = a_y = -9.81\frac{m}{s^2}$ or $-32.2\frac{ft}{s^2}$ if we have changed units from meters to feet (note that 1 m = 3.28 ft). The negative sign here means that the object is falling downward, in the "negative y" direction. If the sign were reversed, say you threw the ball upwards, a_y would have a positive sign. This is a mathematical way of denoting the fact that forces are vectors. The 2-D representation of the acceleration for a basketball being dropped from the scoreboard is:

$$a(t) = (a_x(t), a_y(t)) = (0, -9.81)\frac{m}{s^2}$$

In this example, we have a zero for a_x because the ball is falling purely in the -y direction. If we tweaked the problem slightly and analyzed the ball as it left a shooter's hand, we could still only allow a_x to be 0 if we neglected the frictional force or drag, which the ball feels as it flies through the air.

Example Problem

A basketball player is getting ready to shoot free throws. The ball is given initial coordinates (x, y) = (15, 6) feet, where x is actual distance to the free throw line and y is the initial height of the ball as the player holds it in his hands. The player gives the ball initial velocity $v_i = 30\frac{ft}{s}$ at 45 degrees. Calculate the maximum height the basketball will reach, and at what point it hits the ground.

The position vector at t = 0 was given to us in the problem statement $(x(0), y(0)) = (15,16)$ ft. The initial velocity in vector form looks like:

$$v_i = (30cos(45), 30sin(45)) = (21.21, 21.21)\frac{ft}{s^2}$$

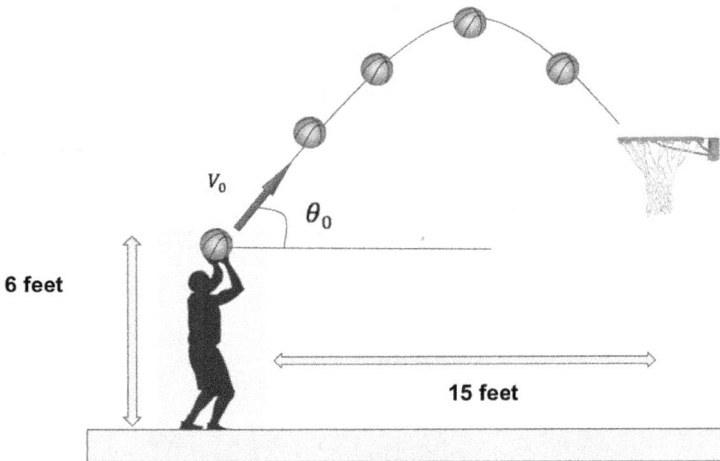

Figure 2.3 Projectile Motion Example

We can simplify equations 2.9-11 by assuming that the horizontal (x-direction) force is zero, so the ball's velocity at all times is given by its initial velocity $v(t)=v_{ix}$. The situation in the vertical direction is only slightly more complicated. The vertical (y-direction) position and velocities as a function of time are given by:

$$y(t) = \frac{1}{2}(-32.2)t^2 + v_{iy}t + y_i$$

$$v_y(t) = (-32.2)t + v_{iy}$$

$$v_{fy}^2 = v_{iy}^2 + 2(-32.2)(y_f - y_i)$$

When the object reaches maximum height, the force of gravity is starting to win over the force that our ball player gave his ball. So at the top of the arc, the ball stops moving vertically for an instant, $v_y(t_{top}) = 0$. We can compute the time at which this happens (t_{top}) with the following:

$$t_{top} = \frac{-v_{iy}}{(-32.2)} = \frac{-21.21}{-32.2} = 0.66s$$

$$y(0.66) = \frac{1}{2}(-32.2)0.66^2 + 21.21 * 0.66 + 6$$
$$= -7.01 + 14.00 + 6 = 12.99 \, ft$$

To find t_f, the final time for the ball to hit the ground, we set the vertical position to zero and solve for the time.

$$0 = \tfrac{1}{2}(-32.2)t^2 + 21.21t + 6$$

That is solved using a quadratic equation. Just in case that's unfamiliar to you, here's a quick overview. The quadratic formula is used to find the solutions to equations of the following form:

$$0 = at^2 + bt + c, \text{ with } t = \frac{-b \pm \sqrt{b^2 - 4ac}}{2a} \qquad [2.12]$$

Using this handy formula, we use the positive root of the solution and obtain $t_f = 1.56$ where $x(1.56) = 21.21 * 1.56 = 33.09ft$. You should note that the solutions we've seen in this problem doesn't allow the shooter to actually make the basket as depicted in the figure. Can you determine how much the initial velocity and angle need to change to make the shot?

Momentum Analysis

When two objects collide, there is a sudden spike in the "contact force" between them. Although contact forces are intuitively easy to grasp (go punch a wall if you don't believe me), they can be difficult to quantify. Rather than use our old friend $\vec{F} = m\vec{a}$ we employ an alternative approach, based on **momentum**. Momentum is a vector quantity of a moving body (or bodies) given by the product of its mass and velocity:

$$\vec{P} = m\vec{v} = \sum m_n v_n \quad [2.13]$$

The second equality shows that if your system is composed of several bodies, the total momentum of the system P is simply the sum of the individual bodies' momenta $m_n v_n$. We generally use momentum calculations to tell us the resultant motion after two (or more) bodies collide. All calculations are based on **the law of conservation of momentum**, which states that the total amount of momentum before or after collisions is conserved (unchanged). In mathematical terms this looks like

$$\vec{p}_{1i} + \vec{p}_{2i} + \cdots + \vec{p}_{ni} = \vec{p}_{1f} + \vec{p}_{2f} + \cdots + \vec{p}_{nf} \quad [2.14]$$

The total initial momentum of a system with n number of bodies is equal to the final momentum of the system. Figure 2.4 shows basketballs of two different sizes (e.g., regulation and youth ball). In each case (a) through (c) the total momentum is the same before and after the collision.

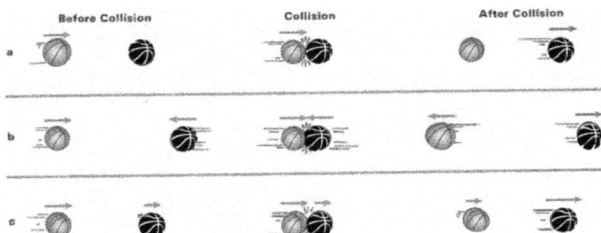

Figure 2.4 Conservation of momentum shown with basketballs of different sizes

Energy Analysis

We may also solve collision problems by analyzing the **energy** of the system, the system's "capacity to do work" in physics parlance. We examine the energy of the system with the **law of conservation of energy**, which states that the total energy of a system is conserved (unchanged). So we can use a given initial energy to determine the final energy of a system, which in turn gives us enough information to calculate other quantities of interest, such as final velocity.

There are several distinct classes of energy. **Kinetic energy** is a property of moving objects. It can be broken down into *linear* or *translational* kinetic energy, quantified by:

$$KE_{trans} = \frac{1}{2}mv^2 \qquad [2.15]$$

where m is mass of the object and v is its velocity. Kinetic energy can also be decomposed into *rotational* or *angular* kinetic energy quantified by:

$$KE_{rot} = \frac{1}{2}I\omega^2 \quad [2.16]$$

where I [$kg\ m^2$] is the *moment of inertia*, analogous to mass for translational motion. Similarly, ω is the rate of change of the angular position is measured in radians per second. This is related to the linear velocity by $v = r\omega$ *where* r is the radius of rotation. Torque $\tau = I\alpha$ is the angular equivalent to the force defined in Newton's second law; here α is the angular acceleration in radians per second squared.

Another type of energy is **potential energy,** which describes the energy that objects have stored in them due to their relative position. For example, a basketball laying on the gym floor has zero potential energy, but a basketball placed on top of the scoreboard has a significant potential energy, which is a function of its height above the ground. We quantify gravitational potential energy with the formula:

$$PE = mgh. \quad [2.17]$$

where m is still the mass of the object, g is gravitational acceleration, and h is the height.

Internal energy represents the energy contained within a system and keeps an account of the energy changes due to a change in the internal state. Such changes are generally due to heat or matter exchange and work, and are in the realm of *thermodynamics*, but we won't cover that here.

Returning to the conservation of energy, the total mechanical energy of a system E is defined by the sum of its potential and kinetic energies.

$$E_{Total} = KE + PE \quad [2.18]$$

Therefore, the conservation of energy states that $E_i = E_f$ which we can write in terms of the energies as:

$$KE_i + PE_i = KE_f + PE_f \quad [2.19]$$

In Figure 2.5 we see that as the height of the basketball rises its potential energy increases while kinetic energy decreases and vice versa. Additionally, it's useful to relate the kinetic energy and momentum using the following relation:

$$KE = \frac{p^2}{2m} \quad [2.20]$$

We'll use this relation in the next chapter for an interesting analysis.

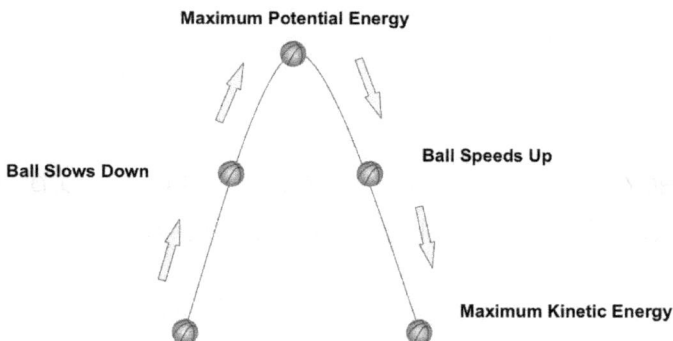

Maximum Potential Energy

Ball Slows Down

Ball Speeds Up

Maximum Kinetic Energy

Figure 2.5 Potential and kinetic energy are conserved as the basketball exhibits parabolic motion

Chapter 3: Physics of Basketball

Introduction

If you are (still) reading this book you probably love the game. I've played basketball consistently since I was a small child and I still can't get enough. And even as a scientist I still sneak in a game a week. My goal is to share a technical analysis of a few aspects of basketball so that students who aren't familiar with the underlying physics principles might connect more with the subject they are learning about in school.

Often, if we understand one thing very well, we can use it as an analogy for many other areas of study. This is something that's done very often in physics and is why textbooks are filled with problems of sliding blocks, springs, and pendulums. Unfortunately, frequently it's hard for students to connect with these examples because they seem rather abstract and uninteresting. It certainly was hard for me to connect! However, I was also exposed to fun projects in physics (more on doing your own projects later), which included designing my own roller coasters and space ships. Because of these, by the time I reached college, I was familiar with some physics principles. With the right examples, you can be too!

Physics teachers may find it useful to start their classes with some experiments that are designed to grab the students' attention as well as help foster a basic understanding. Once we have the students' interest, and the information seems somewhat relevant to them, we have a foundation to build upon.

Those Dreaded Missed Layups!

My first experience on a basketball team dates back to when I was about 9 years old. I can still remember how difficult it was to learn how to make a layup. Our first practice drill every day was the layup line. My coach gave me pretty simple instructions, such as "use the backboard" or "aim for the square." To my chagrin, I'd miss one after the other. When I did occasionally hit one, I tried to memorize the exact spot I'd jumped from on the ground, while aiming at the exact same spot on the backboard. I was hopeless. This experience isn't so different from how some students approach solving physics problems: they try to memorize every little step from some example problem rather than grasping the "feeling" of the material or principles.

Because any hope of becoming a competent player starts with mastering the layup, this seems like a logical place to start applying the physics principles that we have learned in the previous chapter. Ready?

You may remember from our discussion of inertia in Chapter 2 that as we run toward the rim and release the ball, the ball continues forward at the same velocity at which we were running. This small fact is the likely cause for tons of misses. When we release a basketball, the velocity vector has two components: vertical and horizontal, or upward and forward. Many players mistakenly push the ball toward the backboard (or rim) as they release the ball, failing to realize that it's going to travel at that same horizontal speed as your body the moment it's released. Figure 3.1 shows that the additional force toward the basket often causes the ball to have too much speed at impact, and results in a missed bucket. The shooter who is conscious of this fact imparts mostly an upward force on the ball. This optimizes the speed in the horizontal direction, yielding a higher percentage shot.

Figure 3.1 On lay-ups you should push the ball in the vertical direction much more than the horizontal

Expert players have a great deal of touch and feel around the basket. One of my favorite advanced shots is the runner. Those who have mastered this shot may not realize it, but they've also mastered using the concept of inertia. Players are generally moving very quickly towards the basket, often going across the court (sideline to sideline) while stopping on a dime and simultaneously shooting. The player who is successful in these shots softly lofts the ball high the air and manipulates his body while placing his focus opposite to the direction of his movement to compensate for the inertia of the ball. Those who forget about inertia often miss the ball on the far side of the rim.

Mastering Spin

In an era with so many sharp shooters like Steph Curry, I think one lost art (or one which is simply not highlighted as much) is passing. One of the best passers I've ever seen is Jason Williams. Players like Williams have the uncanny ability to fit the ball in tight spaces even when their teammates seem like they aren't open.

If we have any hopes of becoming a great passer (or shooter), we must understand the concept of spin and how it is affected by the ball's interaction with the floor. Spin also affects the way the ball interacts with the rim and backboard—but we aren't talking about shooting just now.

To investigate spin, let's look at the physics of the bounce pass. In particular, let's investigate what happens when we impart different types of spin on the ball as we pass it. Consider three situations: no spin, forward spin, and backspin. In each case the ball is released with the same *translational* velocity and angle to the court, but the *rotational* velocity is different.

In the first instance where there is no spin, the ball comes into contact with the floor and the translational velocity is opposed by the force of friction. Figure 3.2 shows the frictional force in the opposing direction causes a torque about the ball's center of mass, which in turn results in an increased angular velocity in the forward direction. If we remember our conservation laws, then we realize the overall energy has to remain constant. So if the rotational energy increases from zero to some amount, then the translational energy should decrease to compensate for this effect.

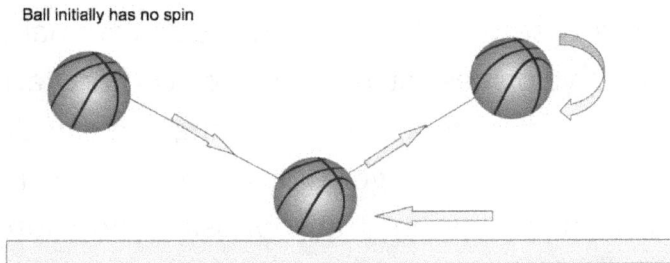

Ball initially has no spin

Figure 3.2 Ball initially hits the floor resulting in a spin in the forward direction

In the case of forward spin, the force at the point of contact points opposite to the frictional force. This results in a decrease in the angular velocity but also increases the translational velocity by an amount which depends on the spin rate. The end result is a ball bouncing forward faster but at a lower height.

Ball with forward spin

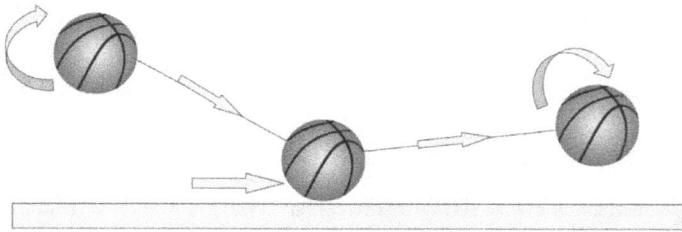

Figure 3.3 Ball with forward spin results in lower trajectory than the no-spin case

For the case with backspin, the translational and angular velocities (at the point of contact) are in the same direction. Therefore, the magnitude of force of friction is larger than in the first example and also points in the opposite direction. A basketball player should realize that if he imparts enough backspin on the ball, it may even travel in the reverse direction upon contact.

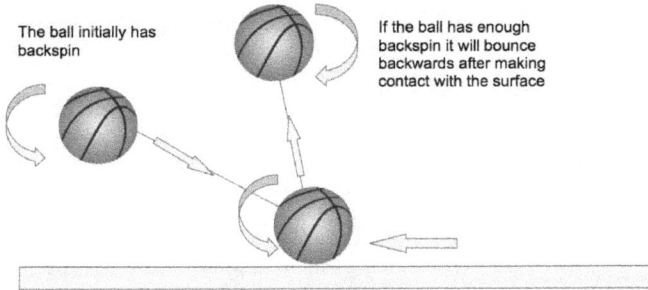

The ball initially has backspin

If the ball has enough backspin it will bounce backwards after making contact with the surface

Figure 3.4 Fall with backspin may bounce in the reverse direction

Let's ground these three situations in some math.

Mathematical Analysis

As mentioned in the previous chapter, it's often most convenient to solve physics problems involving an impact or collision using the conservation of momentum. To do this, we set the linear momentum equal to the angular momentum: $F\Delta t = \Delta(mv_x) = m(v_f - v_i)$ where Δt represents the amount of time the ball is in contact with the court. Now we take the no-skidding criterion (basically the ball rolls smoothly) and analyze the changes in translational kinetic energy ΔKE_{trans} [2.15] and rotational kinetic energy ΔKE_{rot} [2.16], and the total kinetic energy ΔKE_{tot} [2.18]. The first line of the analysis below comes from solving a form of [2.14] for the final velocity:

$$v_f = v_i + \frac{2R(\omega_i - \omega_f)}{5} = v_i + \frac{2R\omega_i}{5} + \frac{2R\omega_f}{5}$$

$$= \left(v_f + \frac{2R\omega_f}{5}\right)\frac{5}{7} = \frac{5}{7}v_i + \frac{2}{7}R\omega_i$$

$$\Delta KE_{trans} = \frac{1}{2}mv_f^2 - \frac{1}{2}mv_i^2$$

$$= \frac{1}{2}m\left(\frac{25}{49}v_i^2 + \frac{20}{49}v_iR\omega_i + \frac{4}{49}R^2\omega_i^2\right.$$

$$\left. - v_i^2\right) = \frac{-2m}{49}(6v_i + R\omega_i)(v_i + R\omega_i)$$

Using $v_f = v_i + 2R(\omega_i - \omega_f)/5$ and the definition of angular velocity $v_f = R\omega_f$, we may solve for the final angular velocity:

$$\omega_f = \frac{5v_i}{3R} + \frac{\omega_i}{3}$$

We insert this ω_f into the equations above to get an expression for the changes in kinetic energy:

$$\Delta KE_{rot} = \frac{1}{2}I\omega_f^2 - \frac{1}{2}I\omega_i^2 = \frac{m}{49}(5v_i + 9\omega_i)(5 - 9\omega_i)\Delta KE_{tot}$$
$$= \Delta KE_{trans} + \Delta KE_{rot} = -m(v - R\omega_i)^2/7$$

We can summarize the above results by stating that we see a greater decrease in translational energy when we shoot a ball with backspin, compared to both the forward- and no-spin conditions. This explains why shots with backspin have a higher probability of going into the basket, because the ball slows down the most upon contact with the rim or the backboard.

We did not discuss sidespin, which can have both positive and negative impacts on different aspects of one's basketball game. For example, shots that have side spin because of an improper release or too much activity from the off-shooting hand tend to have a greatly reduced chance of going in the basket. However, there are many variations of crossovers and passes which take advantage of sidespin to catch the defender off-guard.

Release Point for Long-Range Shots

Now let's switch gears, and use what we have learned about inertial reference frames, Newton's second law, and vectors to study an absolute fundamental; shooting consistently from long range. It's my hope that you can take what you learn in this section and apply it directly to your game.

Basketball coaches always told me to try to time the release of my shot at the highpoint of my jump, and I've seen many videos making this assertion. I actually shot like this for quite some time, but my shot always seemed to hit the front of the rim. The effect became more noticeable as fatigue set in. To study the physics related to the release point, we will analyze three of the conditions shown in Figure 3.5.

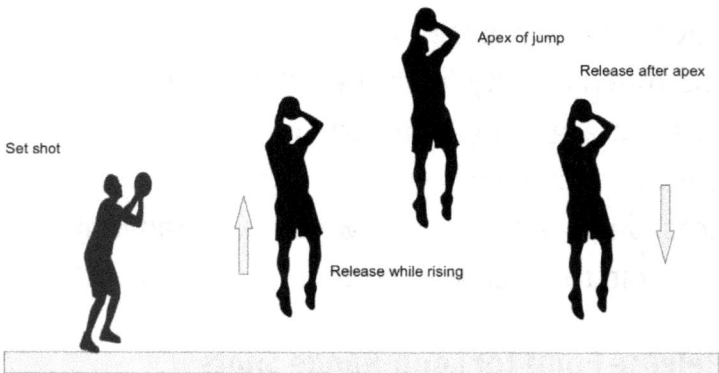

Figure 3.5 Comparing various release points in a jump shot

First, when the individual shooting the ball is stationary and has no velocity (mainly using his arms to power the ball towards the rim). Second, when the player releases the ball as he jumps vertically (before the apex). Third, when the player releases the ball after reaching the peak height of his jump. By the end, you should be able to deduce why releasing the ball at the apex of the jump is much like releasing the ball while on the ground!

To simplify the rest of our calculations, let's ignore the force of gravity (which is the same in all the scenarios anyway) and introduce the following way of writing Newton's second law, so that you can compute the forces easily. Remembering that an object's momentum $\vec{p} = m\vec{v}$ we write its force as:

$$\vec{F} = m\vec{a} = \frac{\Delta \vec{p}^2}{2m}$$

Let's assume that a ball will enter the basket from 25 feet away if it was launched at a velocity of 25 ft/s and an angle of 50°. With this information, we know that the initial horizontal component of the velocity is $v_{ix} = 25cos(50) = 16.07\frac{ft}{s} = 4.90\frac{m}{s}$. Likewise, the initial vertical component of the velocity is $v_{iy} = 25sin(50) = 19.15\frac{ft}{s} = 5.84\frac{m}{s}$. Let's start with the simple example of throwing the ball while standing on the ground. The force you need shoot the ball at the speed defined above is:

$$F_x = \frac{\Delta p_x^2}{2m} = \frac{(0.625kg \times 4.90\frac{m}{s})^2}{2(0.625kg)} = 7.50N, \; F_y = \frac{\Delta p_y^2}{2m} =$$

$$\frac{(0.625kg \times 5.81\frac{m}{s})^2}{2(0.625kg)} = 10.55N$$

in the x- and y- components respectively. In these equations, we used the mass of an actual basketball in kg, converted the velocity to meters per second, and represented the force in Newton's.

Now let's analyze the situation where we are jumping upwards at 2 m/s while releasing the ball. Remember that a ball released will also have the same upward velocity as person who releases it. Because of this, we see that the force in the y-direction becomes

$$F_y = \frac{\Delta p_y^2}{2m} = \frac{m(v_{iy} - v_{jump})}{2m}$$
$$= \frac{(0.625kg \times (5.81m/s - 2m/s))^2}{2(0.625kg)} = 4.54N$$

Of course, the force in the x-direction is unchanged because the jump is purely vertical, so $F_x = 7.50$ N. Similarly, if the shot is released after the jumper's apex, when he is traveling   at -2 m/ s, the vertical force he has to exert to get the ball in the basket is:

$$F_y = \frac{\Delta p_y^2}{2m} = \frac{m(v_{iy} - v_{jump})}{2m}$$

$$= \frac{(0.625kg \times (5.81\frac{m}{s} - (-2\frac{m}{s})))^2}{2(0.625kg)}$$

$$= 19.06N$$

Here again, the horizontal force is unchanged, F_x = 7.50 *N*. Looking at these three scenarios, what have we learned? Your arms have to exert more force if you are releasing the ball at any time other than on the way up. Since your arms are smaller muscles than your legs, it's clear that shooting longer jump shots is greatly affected by the release point. So here's yet another example of what physics can tell us about basketball!

It should now be evident that since the vertical component of velocity is 0 at the peak of your jump, the force to shoot at the apex is comparable to shooting from the floor.

It should be noted that jumping horizontally towards the basket would reduce the amount of arm force needed to make the basket as well. However, strategically, we often don't want to close the gap between the shooter and the defender, so that's only applicable as a special circumstance.

Chapter 4: Philosophy on Self-Improvement (G.A.M.E.)

"G" – Goal Setting

What do you want?

We've talked a little bit about goal setting, but now let's talk a lot about goal setting. It really boils down to being clear about what you want out of life. I've worked with students all over the United States, and I've come to the conclusion that those who come off as having a bad attitude or lacking seriousness are simply those who do not have precise goals to excite them. Even though you might have an idea about what success looks like for you, if you don't have an exact map of how to get to that success, you'll end up going through life trying to get by, and just going through the motions.

If your goals aren't crystal clear and you don't know exactly the type of life that you want to live, then it becomes impossible for you to have the determination, organization, and perseverance required for the level of life mastery that we're talking about in this book. I'm encouraging you to strive for absolute excellence, and a mediocre effort won't allow you to get there. Don't set mediocre goals, either!

First, start off with the question, "What do I want in life?" That's a really difficult question to answer sometimes. But think about it this way; what are one or two things that you'd like to achieve if you were guaranteed success? What are one or two things that you'd like to achieve that are challenging, that will require you to grow, that will make you proud? You might really have to take some time, write down a number of ideas, and really pick what is just out of reach, but may be attainable.

Most of the time it won't be obvious that you can achieve a lofty goal. For example, I work with a lot of students who say, "Hey, Mr. Bohler I want to be a professional athlete." And I'm not the type of person who tells them that it's impossible. It *is* possible; there are plenty of professional basketball players all around the world. The question is, do you have the work ethic?

Are you making the decisions that are consistent with becoming a professional ball player? And 99 percent of the time the answer is no.

Whenever you have a clear example of somebody who has achieved your goal, that goal becomes easier. Things that have already been accomplished by other human beings are always possible. Remind yourself of this when you set really challenging goals for yourself. I want you to grow to strive for the types of goals that haven't been accomplished before. Those are the things that excite people, such as searching for a cure for cancer (yes, physicists are helping with this) or exploring the frontiers of dark matter and energy, or something that's going to have a real impact on humanity. These are the types of goals that inspire other people to get involved in your causes, and their help will really push your forward along your journey.

There's an old saying: Shoot for the moon, at least you'll land among the stars. Though I personally don't use that saying, it does make a very clear point and one that's relevant to our discussion of goals. Striving for a particular goal is very important, not just for achieving that goal, but for *who you have to become to achieve that goal*. If we think about somebody like Steph Curry, who has worked very hard to make it to the NBA, but even when he got into the NBA, you can find stories online detailing how much he practiced at perfecting his craft and how dedicated he was to take his game to the next level. Yes, he's seeing the fruits of his labor, and that is very important, but even if he weren't an MVP basketball player and he just made a marginal improvement, I guarantee you that the discipline he needed in his life in order to get to this level, probably the selflessness, some of the leadership characteristics that he needs to have to be respected by his teammates, are all things that are going to benefit him in the rest of his life. The point I'm making here is that yes, having a goal is important, and yes, you should strive to achieve it. But the person you become on the road to success and achieving a great goal is the true gift of setting important and hard and difficult goals.

SMART Goals

Once you've identified the thing that you want to do in life, it's time to use a well-known framework called the SMART goal system. It's pretty funny, if you look up "SMART" goals you'll see that this is an acronym, but myriad definitions exist for this acronym, depending on the business website or book that you look up. When I say SMART goals, I mean that they need to be Specific, Measurable, Action-oriented, Result-focused, and Time-based. I always like to add Challenging to my particular framework for setting goals. You want to endeavor to do things that push you to improve.

The first key to success is that your goals have to be specific. Your goals have to be detailed and written down. You can't be speaking in generalities. An example of an overly general goal would be, "I want to improve my grades," or, "I want to make more money," or, "I want to improve my jump shot." Your goal has to be concrete enough so that you can clearly visualize and imagine the exact things that have to happen for you to be successful. "I want a 93% or better in biology and math this semester," "I want to earn an extra $5,000 this year," "I want to make 7 out of 10 jump shots." These goals are specific.

The second key to this system is that your goals have to be measurable. What metric can you attach to that goal so that you know whether or not you're achieving it? If we go back to our example of a student in a course, obviously they could use their current class average; an athlete could use points scored per game, or any other statistic that he or she feels would be connected to performance.

The third key on the path towards successful goals is that they must be action-oriented. That gets to the heart of the matter: When you're setting a goal, you have to actually change what you're doing. You have to come up with a list of actions that you must take in order to achieve that goal. It's very important that you identify what you have to do that is different from what you've been doing before.

Moving on to the R: result-focused. Now that you've identified what actions you want to take, what key outcomes or milestones will you use to identify whether or not your goal is being achieved? This is very similar to the measurable, however, the difference is that being focused on results means you will look at that data, track it, and determine whether you are on track. You don't get credit for just trying, you get credit for actually achieving the result. At some point you might have to take a step back and tweak your actions, just to make sure you're on the right track.

The final key to success in achieving our goals is that they have to be time-based. Set yourself a deadline. If they are not time-based, then they're simply a wish. There are no unrealistic goals, only unrealistic deadlines. Once you set a clear schedule for achieving your goal, set a deadline and try to hit it. If you miss that deadline, set another one, and another one, and keep going until you finally meet your goal. But it's a very good practice to always set a little bit less time than you have to achieve certain things as a way of keeping yourself motivated.

Then comes my addendum, "challenging," the one that isn't a part of the acronym. Life is about stretching; it's about improving, getting better. A part of human nature lets us feel fulfilled when we know we're improving, and setting challenging goals is really a way for us to be able to take a look back at a number of things we've accomplished and see that we've actually improved. Don't set goals that you know are easy, set goals that require you to improve, stretch, and dig deep to become a new you. You'll thank yourself later.

Why is Your Goal Important?

The most powerful aspect of goal setting that will allow you to overcome setbacks and keep you from feeling like you want to quit is to identify why this goal is important to you. Your "why" allows you to connect a particular goal with your life's purpose. You can have many "whys" and the more reasons you have for achieving a goal, the more excitement you can muster for it. As an exercise, I'd like you to list at least 20 benefits of your goal. Why is it important to you? What difference will it make in your life? Then see if you can list 20 more—once your goal is met what positive impact will it have on the lives of others? These are the types of questions you should be asking.

Just to give you an example, one of my goals as a scientist is to help thousands of students gain access to technical careers. I could easily list 100 reasons why this is important to me. This list fuels my enthusiasm for unpaid work, to go out and volunteer, and to continually meet and tutor students in my spare time. I do this kind of outreach because I grew up exposed to two different sets of people. I had my friends in the neighborhood, and then I had buddies who were much more academically engaged, like my classmates in magnet schools and college. Since I was a science and engineering major, half of my friends were (and still are!) scientists and engineers. And honestly, my friends who are in

Science, Technology, Engineering, and Mathematics (STEM) fields seem to lead much fuller lives regardless of their occupations.

I've seen so many examples of how the study of science and engineering allows you to have an outstanding lifestyle, regardless of whether you end up in business, law, technology, or the arts. There are so many opportunities in sports for science and engineering majors that I want to push as many people toward this field as possible. On the other side, I've seen students who didn't take school seriously, like many of the guys that lived in my neighborhood. Many made a few wrong choices and ended up facing some serious challenges in life, and at my age they don't seem like they're enjoying themselves as much as they could be.

When I compare those two possible outcomes, and I think about why I'm working so hard to help more students gain access and get excited about learning, that gives me the power to overcome any challenge. All of this has motivated me to achieve my goal of being a professional who is in a position to teach, hire, and create opportunities for those who want the most out of life.

But back to work. After you list your reasons, I want you to list some of your less important goals. I borrowed this idea from Warren Buffett, who says you should make a list of your top 15 goals and then cross out the bottom 10 or the bottom 12. This practice is really important, because minor goals will end up occupying all your time. You will end up too distracted to do the things with most impact for your life. So I'd like you to sit and think about some of your less important goals and make sure you come up with a plan to put those to the side for now, or to accomplish them at a different time in life.

I also want you to list some things in your life that you are willing to exchange for your future success. And these could be something like sports, it could be video games, or spending a lot of time hanging out with friends. When you come up with a plan for success, you have to sit down and do something called counting the cost. You have to know exactly what you are willing to give up in order to get where you want to be.

To give you an example, as a scientist I want to be at the top of my field someday. But, at the same time, I want to be in great physical shape so that I can bring enough energy to my work, and be healthy, and be vibrant. I have access to great fitness centers, and I love to work out; I love to play basketball; I love to run; I love to lift weights. I actually have to limit myself and track how often I'm in the gym, so that I make sure that I'm not taking away from my larger goal, which is to make a bigger impact. So, yes, my health is absolutely important, but I don't want to overdo it, because I can be perfectly healthy without spending two or three hours in the gym every day.

Obstacles

Whenever you set a goal in life, you should also take some time to think about the obstacles that you are likely to face while pursuing your goal. This is important because you want to formulate a bulletproof plan to avoid anything that is going to set you back, or prepare for those difficulties that you know you are going to face. Another area where advance planning becomes crucial is if you don't have the knowledge on how to avoid things that lurk in your future. When you have completed the above, start assembling your team or a group of people who might be able to help you avoid some of these hardships. Let me give you an example, so you understand what I'm referring to.

Let's say you set a goal of being a college basketball player, a D1 athlete. Let's say you want to get a scholarship to college. Think about the common mistakes people have made from a logical point of view. For me as a black man in the United States, some of the risk factors are very clear. Very high on the list is incarceration. It might be teen pregnancy. Low academic success keeps a lot of guys out of college and other D1 programs: poor grades, not even graduating. Lack of curiosity. Drugs or alcohol addiction. Picking the wrong friends. Not being personally accountable and blaming other people. Not having learned how to learn. On the athletics side, you have to do your thing on the court. So maybe it could be lack of exposure to scouting, lack of performance in a game, or injury. I believe there's a certain amount of luck associated with being selected as a D1 athlete.

You can see where I'm coming from. After you've identified what mistakes people make— not working hard enough, not getting a personal coach—you can identify the things you have control over. I'm just going to pick some of the more egregious: jail, teen pregnancy, or drugs and alcohol, these are things that it should be very easy to avoid, but could derail you really quickly. To achieve your theoretical goal of being a D1 athlete, you want to make sure that you're hanging around people who are focused, high-quality, and who have similar goals. Make it a rule to only spend time people who are on the same trajectory as you. Don't drink, don't smoke, don't have unprotected sex. Set yourself these boundaries (think back to what you're willing to cut out to achieve your goal). And then when you share your dreams with your close friends, with your parents, with your coaches, these people can become accountability partners.

Accountability partners are special because these are the people who truly care about you. You've shared your goal, you've shared what obstacles you may face, and now this team of people are going to work with you to make sure that you don't get tripped up by any of these easily avoidable pitfalls.

Your Plan

We've done a lot up to this point. We have identified our goals, we have made sure that they're SMART, we figured out our reasons for having these goals, and we know that we have obstacles in front of us, so we've we realized what we need to do to avoid them. Now here's where it all comes together. I want you to sit down and write a detailed plan of action. Writing your plan is going to require a little bit more prework in this section. Deconstruct your goal. When I say deconstruct your goal, I want you to figure out what are the six most important skills that you currently <u>do not</u> have that you will need to succeed. When we approach new things in life, whether it's going into business or becoming a scientist or playing a sport, it's all the same. Being world-class doesn't require you to master a hundred things, in fact it requires you to master very few. If I were talking about a basketball player, a person who wants to do great on the court, and off, what are the five skills that that person needs?

To continue with the example of basketball skills, a guard has to have good passing skills, be a proficient shooter outside, finish well at the basket, be able to play solid defense, and has to be in excellent physical condition. So those are five major skills that a guard needs to have. And then the sixth might be academic strength so that they could go to college and be noticed by NBA scouts. Now when I look at that goal, I know what I need to be able to do. I know that I need to come up with a plan that's going to allow me to be strong in each one of those areas.

Let's return to our college basketball example. Let's say I'm in the ninth grade and I know I want to become a D1 college athlete. In terms of skill development, you want to be efficient, short range, then mid range, and then long range. It doesn't make any sense to practice shooting a bunch of 3-pointers when you can't make layups or finish at the basket. Let's add major milestones to those skills. An obvious example of a milestone that you need to be able to hit the entire time is maintaining a high grade-point average. Say you decide to set your GPA for 3.5, so that you can go to a school that is outstanding academically and athletically, like Stanford University or somewhere along those lines. Now one of your major steps is improve your GPA, or maintain a high GPA. We focus a lot more on academics in the next chapters.

Another milestone you might want to set as an ninth grader is making the junior varsity team, then improving in all the different skill areas to make the varsity team. Identify what your major milestones are and then constantly come back and measure.

Before this entire plan gets executed, you have to set up a schedule so that you can work at it *daily*. A lot of people can get all the way to this step, but they can't imagine working at their goal *every single day*, and that's literally what it takes. Because every day you're either getting stronger or you're getting weaker. Every moment wasted is something that we could have been doing in order to pursue our dreams. You want to make sure that you set aside time every day that's just for learning. It can be early in the morning, or late at night, but it's focused on learning. And then you want to set aside a time for practice. You may have your team practice for a certain part of the year, so then you have to schedule your individual practice.

"A" – Assessment

Vital Statistics

Now the fun starts. The assessment. We have to figure out how good we have to be to really compete. There are a few different things that you have to do, but I always like to start with the "vital statistics." These key areas of achievement to which you can attach numbers are so important. We've used a couple examples here, such as a basketball player, and we've used a student, so I'll start with an overview of what the vital statistics are for student-athletes. First it's the GPA. Second, for a student headed to college, it's the number of Advanced Placement (AP) and honors courses. Third, it's your standardized test scores, SAT and ACT. Fourth, it would be the number of side projects. When I say side projects I mean independent research. Fifth, your community service hours and involvement. Sixth, extracurricular activities. These six numbers are what colleges use to determine whether or not they want to let you in.

If we were to use a basketball player, their vital stats would be the number of games played and games started. Points per game, field goals made, attempted, and the percentage. Free throws, made, attempted, and the percentage. As well as 3-pointers, made, attempted, and the percentage. Number of rebounds, assists, steals, blocks, turnovers. I could look at all of these vital statistics and with never seeing a person actually play the game I could tell you whether or not he was effective as a player.

The next piece in your research is to know the metrics of the top performers, which is crucial. In terms of GPA, if you want to go to a particular college, then do a bit of research and learn what level of academic success current students are achieving, so that you can make sure that your numbers are equivalent. I'm going to use the example of Stanford University. According to the school's website, 60 percent of incoming freshmen in 2016 had a GPA of 4.0 or greater while only 14 percent of the class was admitted with less than a 3.7. You could also review the breakdown of standardized test scores in the same class. More information might be accessible by calling the admissions office and asking a few questions. You might ask how many AP courses students typically took. Finally, it's a good idea to reach out to a few current students and talk to them about key parts of their application such as admissions essays and community service involvement. You'll be sure to pick up a few tips that aren't even on your radar. At this point in the process you would have more than enough information to judge how you are measuring up as well as some key steps that could be taken to improve your application.

Athletes should make it a point to review recruiting statistics and current seasonal data to determine how well the top players are doing. How many points were they scoring in a game, what shooting percentages did they have, how important was rebounding to people who got recruited to the types of schools that you wanted to go to, was there an emphasis on steals, blocks, and turnovers? It's up to you to do this sort of analysis, so then you can understand the entire landscape related to making it to the next level.

All of this is to help you understand where you are so that you know where to go. Is your GPA on target? Are you on task to take AP classes? If you're a freshman in high school, how was your practice SAT score? Are you on par? Because if you aren't, there are some changes that need to be made. Are you doing any side projects, extracurricular activities, summer programs, or community service? These are all things that you have to start measuring so that you know whether or not so you can adjust your strategy and expectation accordingly.

Self-Testing

Now that you are armed with the vital statistics, it's time to figure out where you need the most help. Self-testing starts with testing yourself on the foundational skills, which must be mastered before moving on to intermediate and advanced skills. What I'm going to do here is give you a preview of what we're going to do in the following chapters, in which we work on a specific set of skills.

Starting with the basics, you need to set up your own drills so that you can understand where your weaknesses are. In terms of math, let's talk about algebra specifically, because to do physics or any other type of science or engineering (and really just life in general), algebra is pretty important. There are some really basic algebra skills that you need to know: integers, exponents, polynomials, and factoring.

A simple way to self-test would be to grab an algebra textbook and give yourself, let's say, five problems from each chapter. Just randomly pick five problems, give yourself a time limit, no notes or anything like that, and sit down and see which ones you can do. And then you go to the back of the book, find the solutions, check your answers, and then tally up and identify which area you scored poorly in. Some people might score very poorly on working with integers. If you have an issue with integers, then it doesn't make any sense to start working on factoring, or getting into some of the other advanced topics that you might even be struggling with in your current class. That's why self-testing is so important. Do you not understand the concept of exponents, is that what's holding you back? Because if you don't know exponents, you can't get into solving some types of equations, which could be your issue when it comes to things like doing word problems. This is how you identify your weak spots.

The same concept applies to playing basketball. Let's just start out with the most basic thing we can think of: at the top of our list of drills is making five layups. We want to make five layups with the left hand, five layups with the right hand. Note your score. Then you want to shoot five free throws. Note your score. Then you want to shoot five mid-range bank shots. Note your score. And then five 3-pointers from your favorite spot. That would be really simple. It doesn't make any sense to move on to shooting 3-pointers if you don't have a high percentage of making free throws, or an easy bank shot. A lot of young people spend so much time practicing the wrong skills. Of course, the game gets a lot more complicated, but you just want to make sure that as you're building your skill set, that you start at this very basic introductory level. Because if you have problems here, you're not going to be able to do well on the more intermediate and the more advanced levels.

On a final note, if you're trying to figure out where to start, make sure that you compare your practice results with your real game results. Take a look at your old exams and see what the common mistakes are. If you're making errors in your algebra problems and you're having a problem with factoring or completing the square, then hopefully you should get a pretty clear indication that this is consistent with your self-testing. Likewise, if you're in a game, you might want to review some of your stats. I don't know how much you'll have them available to you; but where are you making shots and where are you missing shots? Keep a mental track where you have difficulty. And that should definitely be consistent with your personal drills.

"M" – Mastery

"Your" Practice versus "Their" Practice

It's time for mastery. After all the self-tests, we now know where we need to improve. This is where the proverbial rubber hits the road. Now's the time to put in all the hard work that's really going to pay off. I want to explain two key components of mastery, and that's "your" practice versus "their" practice. Your practice will be based on your individual needs (i.e., your self-testing), your individual drills, and your individual assessment. "Their" practice refers to anything that occurs in a group setting. Both are good and very important, but one of these is going to benefit you even more than the other. Can you guess which? You want to be as efficient with "their" practice as you can, but you want to maximize the amount of time focusing on your practice. So how does this play out?

Your Practice

Let me give you an example. As a student, everything that you do for school is their practice. You're sitting in the class, you're listening to a lecture, you're getting homework. These are all generalized things, given out to all students to help them gain a certain amount of knowledge. And then there are all these evaluations: tests, exams, standardized tests, and the like, which are supposed to evaluate whether or not this practice, their practice, was sufficient. We all have to perform on the same evaluations, but your practice is the collection of things that you do on your own to ensure that you meet their milestones.

So, you've taken the assessments, you know where to improve, what should your practice look like? Your practice should focus specifically on where you fall short, and always "go the extra mile," which is the key to personal improvement. The first thing to do is review. Ask yourself, "do I understand the fundamental concept here?" Whether we're talking about a sport or something in class. If you're missing a lot of free-throws, do you understand why? A lot of people don't, and they go and they practice aimlessly for hours, and they don't really get better. This is why it's important to seek the help of others: a coach or a teacher. The same thing can be said for somebody who's having trouble factoring in algebra, or doing a certain type of physics problem. If we don't understand the fundamentals, then we will waste tons of time thinking that we're developing ourselves. Here's the first step, again: clarifying the fundamental concept, which is often done by working with an expert.

If you don't have access to an expert, read a good book. That could be a good textbook with lots of example problems, or a book written by your favorite basketball player. You could also watch videos, as there is so much information available on YouTube. Making use of these resources ensures that you are approaching your goal in the most efficient manner. Why waste time being wrong?

Now that you understand the fundamentals, it's time to practice that fundamental concept in isolation. This is where you set some time aside, and for a week or two you just work on this one skill. If it's factoring, I want you to take 100 problems and break them in the sets of 20, then you do 20 problems, and you continue to work on this one concept until you can score 80% or better in a given time frame. And then you work through your entire list. You go, "integers, okay, I want to move to exponents 80%, polynomials 80%, factoring 80%," and you slowly build up your foundation and your fundamentals over a consistent period of time.

The same with free throws. If you're working on free throws, I want you to be able to make five free throws in a row. From your research, you understand one of the key components to making a free throw shot is to have the right amount of spin and release. If you can't make five free throws in a row, set aside 30 minutes every day to get in the gym and work on those.

As a side note, a lot of you might say: "hey, I've never really been good in math and science in school." The summer is the absolute best time to develop your practice. I'm not talking about a huge chunk of time, you can set aside two hours a day and just grind on your fundamentals until you improve. Then you move up to an intermediate skill, followed by an advanced skill, then you restart the process. The same thing is true in the gym. You start with lay ups. One thing that you'd have to do here is when you're going to try things that are a little more difficult, like 3-pointers, you have to use your personal experience to guide you. Should I be able to make five lay ups in a row with each hand? Yes. Free throws? Yeah. Should I be able to make five bank shots? Yeah. Now, 3-pointers for some might be difficult, so you have to pick a number that fits. But the important thing is that you're going to practice on your key area in isolation, alone, free from distraction.

Their Practice

Their practice refers to standard or non-personalized instruction you receive. This can mean a lecture aimed at all students in your class, not individualized for you. You have homework that's just general and given to everyone, in which everyone is solving the same problems. Their practice can also mean your sports practice, like a basketball practice. Those are more about how the team works together, rather than you improving as an individual. After all, there's really only so much time in a practice. In their practice, you are not the focus, but it is still very important!

Their practice gives you the opportunity to work on things like your personal character, your ability to work with others, your attitude, your curiosity, and your leadership, which are all things that really have a high correlation with success. Their practice—working with other people—is very, very important, and you should not underestimate it. I do think a lot of students take it seriously already, at least in sports. My point in talking about their practice is to remind you to be as efficient with this time as possible.

Efficiency in their practice refers to understanding the main concepts and writing them down so that you can work on them by yourself later, in your practice. I see a lot of students not taking their academic practice seriously: students will sit in a lecture, day after day after day, and feel comfortable just being present. This is a mistake. You have to take "their practice" in the classroom, and see the main points being made, and you have to be able to connect with them, and annotate them in such a way that you can then figure out how to fit them into your toolkit, so that their practice becomes a permanent piece of your practice.

It's the same thing in basketball practice. You might be scrimmaging five-on-five, working on a few plays as your coach is going through the playbook. The focus of this practice is really on the team, so you want to make sure your takeaways are whatever you need for synergy with your teammates, and memorizing the plays so that you know the playbook. And then your coach will give you different keys, things that you should work on, such as your ability to see the entire court, or pass the ball, or even just the mental part of the game, so you are more consistent. After every practice sit down and take a look and think, "what can I take away from that, what was set for me, what did I learn?" and then take some time by yourself so you can put it in your toolkit, "your" practice.

Their practice gives you great quick feedback and then leads you in the right direction, and helps you get the big picture. But you do most of your individual improvement and your personalized practice that you schedule and design yourself. This is really where a lot of people fall short. They just do the minimum. They just show up, go through the motions. So, I know that if you follow my process to a 'T' and dedicate yourself to your practice, and use their practice efficiently, you will smoke your competition every time. Other people just don't follow through, but you do.

"E" = Engraining

Increase Resistance

Engraining (ingraining) is where we solidify our mastery, so that we're able to transfer our new skills from the practice court into the real game, or we're able to take what we learn from studying and use it to ace a test. In order to hit this phase, you need to identify what your limits are—you've been through all the other steps—and then practice just beyond your limits to make sure that you're still stretching yourself.

One of the logical ways to do this is to choose more challenging drills. We talked about previously the basic set of jump shot-related drills. Now instead of hitting set shots, repeat all the same drills but dribble really hard to a spot and shoot. Capture the numbers and the statistics and take note of your drop off in performance. You want to do drills after you've had "their" practice or group practice and then practice again while you're fatigued, and see how that affects stats.

Another variation is catch-and-shoot drills. You'll need a partner. You want to run to the spot, receive a pass, and see how well you can prepare for the shot before you shoot, thereby catching the ball and hitting the shot all in one smooth motion.

The reason why these types of things are important is they simulate what happens in an actual basketball game. For that reason, I love the idea of going to the court and adding obstacles, like chairs, or different cones, or even using a technique where you have to pick up small tennis balls or other objects as you dribble around and hit different shots. It's a way to add just a little variability that forces your body to adapt to a nonstandard circumstance, which is something that's really going to be beneficial to you as you're trying to improve.

For studying, if it's math, physics or some other technical subject, what can you do after you've done your homework and understood it, and you think you've mastered these concepts? You could try more difficult problems. Many classes spend time focusing on the most basic problems, which are often of the "plug and chug" variety that just require you to plug numbers into a formula. But these are not the type of problems that are curve balls, or the types of problems that we really get stuck on during a tough exam. Instead, we want to find tough problems to solve on our own so that we can get more comfortable dealing with these sorts of things.

Another example for increasing academic resistance would be to read more difficult texts. The example problems that you see in an introductory text, be it an undergraduate text or a high school text, present the materials one way, and then as you go up a level of expertise, the information gets presented in a different way. So even if you're just reading through the examples and making sure that you understand what's going on, that will also solidify what you've been exposed to, or at least let you formulate some new questions.

A piece that I don't see often enough, but something that was really key to fostering my love for physics and engineering, is to conduct your own experiments or to play with do-it-yourself toys, which require you to build them yourself. That's really my favorite suggestion, because that's the way I got to "play" with different subjects. I was fascinated because I was coming up with my own problems and solutions because I was doing my own thing. Every so often I'd hit a road block and have to reach out for help. This was always a much more efficient learning mechanism for me.

So let's say you're taking a science course, you should consider looking online for some of your own experiments to do. When I was a child, I always looked forward to the science fair, because it gave me an opportunity to choose my own problem, to do my own research. If I've ever been a standout it was in these types of competitions. These were always so enjoyable because unlike class, I wasn't doing a bunch of problems that someone else had already solved. I actually remember the gory details of these projects the most and they happened more than 20 years ago. I shared this anecdote not to illustrate my intelligence but moreso to emphasize that the deepest level of learning occurs when you are doing your own research and overcoming challenges of your own design.

There's one final level to increasing the pressure, and that's adding constraints. When I say constraints, I mean adding a time requirement, or adding repeatability requirements, and then removing helpful resources. Let's go back to our example of a shooting drill. In a shooting drill, now instead of hitting set jump shots or running to a spot or catch and shoot, let's see how many you can hit in 60 seconds. And then you find your benchmark and see if you can do better, and do better, and do better, and you can track your progress over a long amount of time so you can see how you're actually developing. Or you can do something that is even tougher, and that's repeatability. Pick a few things, like needing to hit five free throws in a row, five jump shots in a row, and then five 3-pointers in a row, and you need to complete each level to move to the next, and you're not going to leave the gym until you've completed them all. Sometimes that can take 10 minutes, or sometimes it can take two hours, and I'm telling you, that's a way where you can simulate the types of pressure in game time situations to just test it and practice your nerves.

A great way to make your study a bit tougher is to limit the amount of resources you use to solve sample problems. Since you've dug deep into a particular subject, now is the time to close all books, turn the Internet off, no phone, no help, no friends, and try to do problems just off the top of your head, even doing the math mentally. The only way to improve when performing under pressure is to practice under the same conditions.

Contrary to my prior suggestion, I actually like to use a study group once I feel like I've really mastered a topic. This isn't simply for me ask questions but to be the expert who is able to answer questions and help others. Sometimes you think you really understand something well, until someone asks a question in a way that you haven't thought about it. Or, until explaining some concept forces you to think about things a little differently. It's a great way to study that isn't boring and a great way to review. And the best thing is you get to hang out with your friends and help other people. Study groups are a really fun way to finish your studying and switch things up.

My last suggestion for the athlete is to compete against better people at every given opportunity. In your practice, whether you're playing pickup games or just having fun, compete against the best. That's the way you do it. Don't compete against people on your level, because it limits your growth. When you compete against someone who is a lot better than you, it forces you to dig down deep. Swallow your pride, and learn from someone.

Trust Your Preparation

I want you to visualize success. It's the final road. You're getting ready to compete, and you've put in the work. This is where a lot of people panic, because they don't trust their practice. We trust our practice because we put in more work than anybody. And if you didn't, then you need to go back and check out the rest of this book, because working harder and smarter than others is one of the central themes. There are no shortcuts. As you stare down your game or your test, I want you to visualize your success. Take a minute and actually go through the game or the test, and see yourself succeeding. I want you to remove all limiting beliefs. Just know that if you've done it once, if you aced a test once, you can ace this one. If you've hit a big shot once, you can hit a big shot in a big moment. I want you to make sure that you have a strategy; you've done your research; you know what to expect; you've prepared; you know your opponent. I want you to rest well, eat well, show up early, and win.

Chapter 5: G.A.M.E. for Basketball Skills

We spent Chapter 4 discussing my improvement philosophy. You set goals, you assess your skills, you master those skills, and then you engrain what you've learned so that you can use your new skills to accomplish your goal. In this chapter, we're going to apply my philosophy specifically to improving your basketball skills.

Goal Setting

As you know by now, the first thing you want to do is set a very specific goal. We covered prework, goal setting, and writing down a plan in the previous chapter so at this point you should have a good idea about how to set goals. As I mentioned, it can't be just "to get better," your goal should be something very specific. Try something like, "I want to improve my ability to score," and then refine to, "I want to improve my field goal percentage from 33% to 43% next season." As you improve your scoring percentage your scoring efficiency improves, so overall you will probably see a boost in the amount of points you score.

Assessment

The first step of this goal setting is an assessment. I'm just going to share one drill in particular with you. I'm not a basketball Ph.D. but I am a scientist, so I've approached this suggestion methodically. I've played on several teams and I've done my own research using a ton of books, but there are tons of resources out there for you to do your own research. What I'm doing in the following section is telling you how to assess yourself based on the outcome of this particular drill.

One of my favorite drills is the step-back drill as shown in Figure 4.1. You start by separating the court into five different zones: you have a zone on the left, one in the left-middle, one in the middle, one in the right-middle, and one on the right. The basic premise of this drill is you take a shot from the first spot (closest to the basket), step back, take a shot from the second spot, step back, take a shot from the third spot, and so on all the way out to the 3-point line, and then you switch zones. You repeat this drill for each zone, all the way around the court, making sure to keep track of the number of shots that you've missed and in which zones they were. This set of simple shots is just made to judge your ability to shoot the basketball. Overall, this is a pretty straightforward drill. Although you can do this drill by yourself because you don't need multiple balls or anything like that, the drill is obviously easier to carry out if someone is present to help you out or if you use a rebounder.

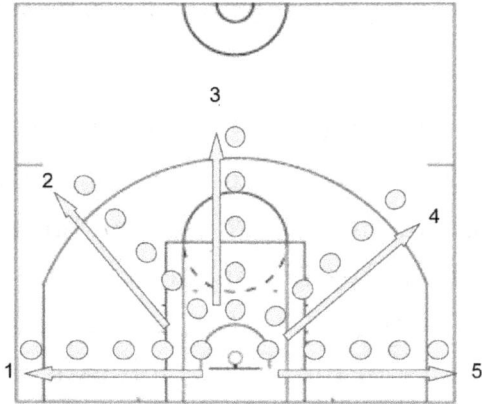

Figure 4.1 Schematic for step-back drill

After you finish this drill and tally up your score, you will probably notice that it's easier to shoot the ball closer in. You may use a chart similar to Figure 4.2 to track your results. Jump shooters know that the best way to improve your ability to shoot is to work from the inside out, so remember this as you set goals. And then, as you tally your scores for each zone, you might notice a deficiency, let's say, on either the left or the right side of the rim. You really want work on this because for your success in basketball, you want to be able to score from both sides of the basket.

Another number that you want to make sure that

you know is the percentage of overall baskets you made. Calculate the total number of misses. So

	Row 1 (Left)	Row 2 (Lt Ctr)	Row 3 (Center)	Row 4 (Rt Ctr)	Row 5 (Right)	Total
Location 1 (closest)	1	1	1	1	1	5
Location 2	1	1	1	1	0	4
Location 3	1	1	1	1	0	4
Location 4	1	0	1	0	1	3
Location 5 (farthest)	0	1	1	0	0	2
Total	4	4	5	3	2	
Grand Total	18	Misses	25	Percentage	72%	

Figure 4.2 Sample tables for results of the step-back drill

what overall percentage of shots are you making? Are you making 20%, making 30%? Depending on your age, strength and skill level, those numbers may be pretty low. In a really fundamental drill like this, you should be able to at least get to the 60% range.

Beyond overall percentage of shots that you're making, look at your scores and see if there's an average distance where you start to see a fall off. Examine your totals from zones one, two, three, four, and five separately, and see if there's a different fall-off distance in a particular zone. This exercise might let you know what range of shot you need to work on as you build up your overall skill set.

Shooting Mastery

Let's talk about mastery. You've assessed your performance on the step-back drill, and you can see there's some work you have to do. To accomplish any goal, you have to make a schedule. For success, we're going to schedule practice time every single day. That's just a rule. If you want to improve in anything, you have to practice it consistently, even if the time is short. You also have to figure out your daily baseline in terms of practice workload. In general, go into your practice knowing how many shots you're shooting that day. This is critical, even more than improving your shooting deficiencies as pointed out by the step-back assessment. Basketball is a tough sport, and the more shots you shoot everyday, the better you're going to get. That's really one of the fundamental laws of shooting.

Decide in advance what your daily shooting number will look like: Is it one hundred? Is it two hundred? If you're really serious, you at least want to be in the 400-500 shot range. I've even seen some shooters who schedule 1,000 shots a day. Perhaps not all in one session, but some might. Your elite professional players certainly might set a goal for 1,000 makes in a day, or even 2,000 shots, which might take several hours depending on the skill level. This is the level of planning you need to bring to your schedule—maybe not 1,000 shots tomorrow, but plan out how many shots or makes you will take.

Another thing that should be a part of your daily plan is improving your conditioning. Identify your workout schedule; if you don't have one, you should work with your coach to establish how many sprints you should be doing and how much weight training. Ask about running some long distances, too. From a physical perspective, all of these things improve your shot, and many other parts of your game too.

The last really important thing to note here is technique. I've mentioned technique before, but I'm doing it again to make sure that you have some clarity about the best approach. Make sure you spend a little bit of time understanding the reason why your performance is wherever it's at— because your technique isn't perfect. If you spend thousands of hours shooting with poor technique, you will still get better, but not as fast as you might if you made that correction to your technique. You want to work with your skills coach, your basketball coach or an outside skills coach. You want to do things like study game films. YouTube is a gift; it gives you access to watching the all-time greats in slow motion. You can also record your personal jump shot and compare what these other individuals are doing on YouTube.

As a side note, everyone is unique, and you have to compensate for your own body. That being said, having awareness of some fundamentals in your jump shot is important: where you keep your eye on the rim, where your shoulders are, where your feet are set, and making sure your elbow is in front and not flailing to the side.

So we've taken a look at your schedule, and said that the more you shoot, the better you'll get. Now you realize that at every practice, I want you to track your total number of shots made and missed. This is something you can just write down in a practice journal, or keep on your phone. Remember, you want to keep a results-based approach (think SMART goals!), which is really useful in case you practice for a few weeks and aren't sure whether you're improving.

To work on your basketball-specific deficiencies, identify a set of drills to help you. Like I said above, learn to shoot from the inside-out. A useful and common drill is to go directly under the basket and make 50 shots on the left, on the right, and in front. Do this every time you step in the gym, and you make those 150 shots with nothing but net and a swish. The reason why this is such a good drill is because you're so far under the basket that you force the ball to have the right type of arc. You can't shoot flat. In other words, this is just a great way to work on your touch around the basket.

As you start to extend your range, you could work on the step-back drill that we did in your assessment. But now before you move on to each spot, you have to make five shots. And you can increase the level of difficulty by having to make a certain amount of shots in a row depending on your skill level. Now that you have your plan for practice, you're practicing every day, you're judging how much better you're getting, you have some outside help, the question is how to transfer.

Engrain Skills by Increasing Difficulty

In Chapter 4, I talked in general terms about transferring skills from practice to success in real life so that you can accomplish your goal. The question facing us now is how to transfer all of this practice in the gym to improvement in your game. As before, the main way to get this skill transfer is to increase the difficulty in your skills practice, testing yourself past the point of what you're actually going to need. This will take the form of variations on what you're already doing in your drills.

If you were doing the step-back drill, you might add something called a spin-out. If you're not working with a partner, a spin-out is when you spin the ball to yourself and then cut to the spot that you're shooting from. So you spin the ball to yourself, cut to the spot and shoot. Spin the ball to yourself, cut to the second spot, shoot, and so on. This would probably be easier working with a partner. Likewise, you want to add all the moves that you're going to do in the game. A lot of times you're shooting off of the dribble, so you might want to spin to the spot, one hard dribble, shot, go to the next spot, two hard dribbles, shot, next spot, three hard dribbles, shot. All of these examples are just adding a mix, so that you get really comfortable incorporating your new shooting skills with the rest of your game.

Of course, there are a bunch of other moves that you want to make sure you have: shooting with an up-fake, a double up-fake, also shooting off of a crossover. And that crossover comes in so many varieties, the possibilities are mind-boggling! You can shoot after crossing over with a step back, or sideways. You know that all of your favorite NBA players have mastered these skills. But these skills have to be practiced independently so that you can incorporate them into your game.

One of the other ways to work on ingraining your new skills is to play against tougher competition. Tougher competition means playing people who are better than you, and you should strive to do that as much as you can. You may also want to play a number of different people in a game such as 21 (one of my favorite games growing up), because it allows you to compete against multiple players simultaneously. You might play one versus two, one versus three, even one versus four. Having the strength to remain comfortable while multiple people are guarding you is very important. Even in your normal practice, particularly group practices, you will cycle through working on your one-on-one skills or your three-on-three skills or your five-on-five skills, and these are all scenarios that will prove very important for success.

Finally, once you've improved your game and worked at it until it becomes muscle memory, make sure that you're prepared to compete. A lot of this part of the process is mental. You have to trust the fact that you did the work. Prepare hard, then trust your preparation. Just to make sure you studied your game plan, your coach has probably set forth some plays. See if anyone is scouting. You should be familiar with as much as you can before you go to the game. And show up well rested, well fed, ready to go, and visualizing your individual success as well as your team success. Setbacks happen, but the better your preparations, the better your luck.

Chapter 6: G.A.M.E. for Academics

Goals

I see too many students going through life with a poor attitude, who aren't striving to be successful. As we discussed in Chapters 1 and 4, oftentimes a student's poor attitude really comes down to their not having a clear sense of what they want to accomplish in life. They haven't taken time to sit and think about the kind of life they want to live and why. Most students haven't seriously thought about what they want to be remembered for or with whom they'd like to spend their time. Once we think about the ideal life as we see it today we can aim to get there. Remember our technique for setting SMART goals: they should be specific, measurable, achievable, result-oriented, and time-bound. I also want your goals to be challenging. Set goals in different areas of your life, whether it's education, health, family-related, etc., and make sure you revisit them daily.

In addition to being clear about what you want, you also really have to believe that you can achieve it. Oftentimes people can't see a goal clearly because their current situation is so dire. One thing that helps me in moments of crisis is to think about other people who have achieved my goal. The great motivational speaker Les Brown even mentions this: If somebody has done it somewhere, then we know it's possible for us.

Our first step is of course to create a goal, one that gets us so excited that it inspires us toward a level of effort that we haven't previously shown. When I'm talking about goals, I don't mean things that are cliché, like getting a 4.0. In five years, few people will care about your GPA. A stronger goal would be becoming an engineer who will build something to positively affect mankind. You should be able to connect every activity you undertake to that worthy goal. Then as you matriculate through an engineering program, you learn as much as possible, give maximum efforts in your courses, and join student-led engineering groups to take on side projects. Of course if you achieve your main goal then you'd likely have a high GPA because you have your priorities well established.

Personally, I like to share my goals and plans with my close confidants who can actually help me! In your plan, list the things that make your goals important to you. As you embark on academic success, always remember your personal "why." Why is this goal so important to you? Why do you have to do well in school? Personally, I think one must strive to do well in school because this is where you "learn how to learn." And picking up new skills and being able teaching oneself is a skill within itself that could lead to your financial freedom.

The next step in this plan is to write down a starting point. Understand where you are now, and then the first steps you need to take. As I told you earlier that SMART goals are time-bound, you have to set a deadline for your improvement. Then you have to set up smaller deadlines, making your progress granular, so you can check on and measure your progress.

Another step in your plan is to identify your real obstacles; therefore, you want to take a look at how you've done in past courses and write down some of your real obstacles. If you have weaknesses, take a second to think about how you can strengthen them. For example, if I'm in a physics class and fail a test early in the semester because of some deficiency in algebra, I have to be able to identify what additional algebra skills I need to develop and then figure out how to improve quickly so that I can pass the next test. Getting over tough starts is something we have to be comfortable with because we don't start off anything new in life being an expert. Having the ability to relax, collect yourself, identify the problem, and address it is a total game changer when it comes to your personal development.

As you think further about obstacles and weaknesses relating to your goals, it's important to understand whose help you are going to need. Let your supporters know what you're trying to accomplish, and reach out to them to come up with a strategy and a schedule to strengthen your self or side step obstacles altogether. Afterward, put it all together, write all these things down, and revisit your plan every day just for a bit. Read it, tweak it, and spend a little bit of time visualizing it. Review your life plan often specially when you don't feel motivated. This will be the key to allowing you to make better and stronger decisions, because it's clear what you want to do.

Academic Prework – 99% of You Won't Do This!

After you have your goals in place, the next step is to have an understanding of your goal's requirements. In academics, this translates to a certain amount of prework that will be associated with passing any class that you have. Prework before the class starts is a process that I think most readers of this text may be unwilling to do, but it completely pays off every time.

The first step is to, in advance, contact the instructor, and actually be prepared before the class starts. This is something that's quite common in some demographics and isn't common in others. There are many students in Silicon Valley who study all summer in order to be prepared to do their best in the fall for particular courses. That's an extreme, but what I'm suggesting here is to understand exactly what will be presented in the course, to know what textbook is going to be used. Get the text early and just skim it. Read it quickly, looking at the titles, the subtitles, the chapter summaries, and the captions. Spend a decent amount of time doing this, so that once the information is formally presented to you in lecture, you're not seeing it for the first time but have some prior knowledge to connect it to.

That's a major piece of academic success that most students aren't prepared to work for; but taking a look ahead and preparing for the future by doing a little bit of work over the summer, or over a winter break ALWAYS pays off.

Homework Is the Assessment

Homework is really an exercise of time management. It's probably one of the most stressful aspects of being a student, whether you are an athlete or involved with other extracurricular activities. But the simple exercise of completing all your work is the basis for academic success. You have to break up your available time for studying into two sections. Content mastery, which I'll call studying, and then actual homework, as in assignments. Our goal for completing homework is to get it done in the most efficient way possible.

I've seen students waste so much time here that it's not even funny. There are a bunch of different examples that I could give you, but I'll just stick with one. In a lot of classes in college, engineering, physics, what have you, homework is worth 10% of your entire grade. Yet, students spend 80% of their available time completing homework assignments and only 20% of their time making sure that they've mastered all of the content. Which is ridiculous! Budgeting your time is important. Figuring out how much time you have to spend on each assignment an putting it on a schedule should keep your life organized and manageable. Here's my simple approach to completing homework assignments on time.

One, we're assuming that you understood what was happening in class. This is possible if you've done your prework, gone to class, and taken measures to make sure you don't fall behind. When you are given a homework assignment, read through all of the problems before you start and give each problem a try until you get stuck.

The one thing you don't want to do is cheat. Students think looking up solutions on the Internet is harmless, but it affects them negatively on so many levels. To put it kindly, you're just missing the entire point of education, and though you might get a short-term grade, it makes it more difficult for you to prepare for the exam. You don't understand what you don't know until you reach the failure point, and even if you do get a passing mark in the class, you didn't learn as much as you could have. And this will be probably the only time in your life that you can devote 60-70 percent of your time to your education, so take it seriously.

Don't cheat and don't worry about other people doing it.

I do realize that many times you might not have an idea about how to do a particular problem. Write down what you do know, take the problem as far as you can take it even if it's the wrong idea. We don't stress out about having the incorrect answer at first, because we know we'll learn from each incorrect attempt. If you get stuck on a problem, put the homework down and spend some time studying: review your class notes, skim through the book for a better understanding, and then go back to the problems. If you still don't know what to do at this point, this is when you reach out for help.

One of the most powerful things you can do is form a study group. In a well-structured study group, you learn from others and teach other people too. It's sort of a flow, there's a budget in which you receive help, but make sure that you are making a contribution. It really pays to balance your study group with one or two people who are smarter than you are, and then one or two people who are struggling in the class more than you are.

You get a lot of out of teaching others and a study group gives you the opportunity to do that.

Everyone in the group should try all of the problems before meeting then you may discuss them and share what you're thinking. Once you have feedback you can work together to come up with a solution. Be careful that your efforts don't overstep the bounds of cheating and make sure that you don't copy anyone's solution. Check your instructor's policies for working with others, but meeting with a group of students who are serious can help you seriously improve your grades and reduce your time spent doing homework.

If you don't have a study group then you should seek other help. You could arrange for help at your school's help center, with a tutor, or with your instructor, but speak up and get the help that you need. If for some reason the help isn't available, make sure you reach out through your parents. The main thing is in order to make the help useful, you have to get to a point of failure, just like with lifting weights. This is what I'm calling the assessment; you don't know what you need help with until you actually get into it and reach the edge of your understanding. But what I *don't* want you to do is spend five hours doing one problem and not getting anywhere. You really don't have time for that. If you think about your day in school, class until 3 p.m., practice for a few hours, another extracurricular activity, eating, you really want to get the homework done as fast as possible so that you can actually set up time to study.

Another Assessment: In Class

After homework, the next critical component for improving your grades and class performance is to focus on what you're doing during the lecture. You either did or did not do the prework. That's fine, you can still survive, but you have to be prepared and present during class. There are so many students who sit there throughout the entire course of a class missing 80-90 percent of the material that's presented. Even though they're following along, writing down everything that the instructor is saying, that information is going in one ear and out the other.

It's much easier to zone out when hearing something for the first time. I often had classes which felt too long. Take a break during the lecture and ask yourself periodically what the main point is. Break down what's been happening in the class to make sure you're pulling out the key points. Raise your hand and ask an instructor to double check your understanding.

To show you how to make effective use of your classroom time, I'll use physics lectures as an example. A few things happen during a physics lecture. The instructor is going to give you some background information, he or she is going to give you a derivation, or is mathematically going to show you the physical principles that are used to arrive at a particular law of nature. Derivations are usually the most difficult part of a lecture to understand. So when those are happening in lecture, you need to understand what key assumptions were made, what the key mathematics steps are to arrive at the final form of the physical law. You can always raise your hand and courteously ask the instructor to clarify a point for you, so that you walk away ready to apply the key points to your own work.

The next thing that happens in every physics lecture is the example. In an example, an instructor is going to show you how to work out a particular problem. As you're listening to the example and you're lost at a step, you might want to ask, "What happened just here?" and, "What are the typical variations that occur in similar problems?" Examples give a good instructor an opportunity to give you hints on topics that aren't explained so well in the textbook. You should write it all down so that you have it available for later use. Examples demonstrate how this problem can be changed, how it typically shows up, and what things are important to be able to maneuver. Pay attention to examples, because many instructors give examples that are the key to doing well on the exams.

In case you haven't figured it out yet, examples are the most important part of the lecture for your success. You should understand every example problem that's presented, but then also ask the instructor (either in class or afterward) about the key variations. You want to take a look at the example problem and identify what to do at every decision point in solving that problem.

I'll digress here a little bit. When I was in school, despite working hard, I had some mediocre semesters. About half of the time I did well and would earn somewhere near a 3.5, and other times I might get a score lower. One day I talked to a friend—I think he got a single "B" during his entire time at college—and asked him, "What's the key?" He told me that the major key to his success is to understand what's happening in class every day and never fall behind. As I think back to all the times where I had really gotten in trouble, it's where I went lecture after lecture, sort of being lost, caught in a cycle of trying to keep up with homework, getting tutoring, all while collecting more and more "intellectual debt" from not being able to understand what happened in that day's lecture.

So your number one priority as a student, even if you didn't do any of the prework, even if your homework isn't completed, is to make sure you keep up with lecture and that you understand *everything* presented in class before attending the next lecture.

To do this, once you get out of lecture, you need to review it, make sure you got the key points, ask yourself questions, and meet with your instructor, friend, or a teaching assistant, whomever can shed light, so that you figure out what happened. Once that has happened you say, "OK, I got it, this is what the focus was." Then you can proceed and do your homework. In fact, this process serves you very well for doing the homework. To fully understand the lecture, you might need to read a bit of the book, so let's discuss power reading while we're at it.

Power Reading

I see so many students who spend so much time reading the textbook one word at a time, trying to glean every little kernel of information that they can, only to realize that they ended up not learning much at all.

In power reading, you want to skim each book (you should try to have at least three texts on each subject) just to see the differences in how they present the information. You don't have to read every word of every chapter but focus on the main points. A lot of physics books are available for free on the Internet, so check out their examples as you skim the chapter. You might even notice that homework problems that are presented in one book are actually very similar to example problems that are given in other texts. So it's very important that you have multiple textbooks at your disposal.

Often, you can see which textbooks to use in the supplemental reading section in your course syllabus. You can also reach out to your instructor and ask, "Hey, I'd like to have another couple of books to use as resources, which ones are good?" You could ask your librarian or just Google some good texts as well.

Another approach to improving your reading efficiency is to circle the most important words in each sentence. In most books 60 percent of the words are phrases that don't add any value, so circle all the words that add value and then make sure you understand the main point. You may even rewrite the sentence only using the important terminology. This method ties in really well with creating concept maps which we shall discuss later.

Formula Anxiety

Formula anxiety is a concept that I haven't seen written about anywhere else, but I know it occurs when people who are less comfortable with mathematics flip to a page filled with formulas. When you don't know what's going on, the first thing you think is, "I'm not going to be able to do this!" It doesn't just have to apply to formulas, infographics, or a chart that just looks completely incomprehensible. This anxiety is a panic that will catch you off-guard and eliminate the right state of mind for studying a tough subject.

I want you to be aware when you have formula anxiety. When that panic happens to you, I want you to immediately close the book or take a break, take a step back and think about your goals. We might not talk about meditation, but take a few deep breaths and remember your goals. Think about why are you studying, and what it is that you want to accomplish. Then think to yourself, "If someone has written it in this book, it has been mastered, and it's possible that I, too, can master it." You have to realize that at some level you may not be able to do a bunch of integrals or derivatives, TODAY. "I totally do not understand how that works TODAY, but I can chip away at it. And because this thing is written in this book by someone else, there's evidence I, too, can grasp what's going on." And you have to know that without a shadow of a doubt. Once you've taken that break, and you've made sure that you have the right state of mind, then you go back in.

Step one to understanding formulas is to make sure you understand the meaning of each symbol. You want to search the text, start with the topmost formula, and make sure you know what X is, what Y is, and so on.

Then you'll think, "OK, I see a different symbol. What do these things tell me?" What does this symbol mean in the physical reality of the system we're examining? Let me give you an example. I know that an integral is area under a function, so if I see an integral in a problem, I have a strong inclination that area is involved somehow.

So now I understand what all the symbols in the formula are, right? I know what all the variables stand for, I understand what the units are, and so on. Then I figure out how to get from the left side of the equation to the right. And this is probably one of the most pernicious steps because a lot of times it's not obvious, and especially when you're reading a textbook, it might be hard to follow the exact steps. So I take a look, and try to see if I can get it when I'm following it. If I don't, circle that formula, highlight it, write it down, and move on. I might find an example where I can see it done a bit differently in another book, or they might show a little more detail. It could be one of those questions I ask in class the next day, saying, "I was looking at this, I understand A, B and C, I missed this one point, could you help me?" That's a really big thing but I have to be able to tackle formula anxiety and overcome it.

Getting Over Sleepiness

Inevitably, when you study you'll get sleepy, or you'll get distracted, or you'll become disinterested. One of the keys to being a very good student is to be able to redirect that energy and refocus. Instead of putting your head down and going to sleep, or completely giving up what you're doing because you realize that you're not interested in it, I want you to do a couple of things. First, I want you to get up and spend five minutes doing something vigorous, whether it's push-ups, sit-ups, dribbling a basketball, or doing anything that you completely love to do. Take 5 or even 10 minutes and go do that thing, and you'll notice immediately that your body will wake up like you've hit the reset button.

After you've hit the reset button, it's a really good idea to change your approach. One often gets sleepy by just rote reading of some text without engaging content presented. If that's happening to you, why not start working on a concept map for the chapter, to make sure you understand the big picture? Or you might want to switch problems and do something else. Is this an assignment that's due immediately, code red? Or perhaps you're getting disinterested because you hit a roadblock, you're banging your head against the problem, you can't quite get it, and your body shuts down? Pick up the phone and find a study group or somebody who's working on that same problem.

This brings me to another point: Do not wait until the last minute! Often students are busy cramming for the exam or staying up all night long to get a homework assignment done. Because of that, they can't take advantage of one amazing feature of the human body: the subconscious. As soon as you get an assignment you want to read through that assignment just so your subconscious can start to work on it. So if you get tired while you're studying and you have time, then it's okay to take a short nap. I find that I often figure out solutions to problems in my sleep. My subconscious gets an opportunity to try the problem and to think about different approaches. And oftentimes I'll wake up and I'll have an idea about another way to solve it.

Content Mastery

In terms of content mastery, what are we talking about here? This is very simple: do you *understand* the key points? You sat in class, so you should at least be familiar with the key points, and after your homework has been graded you realize that there are a few concepts that weren't as clear as you thought. Now you make sure you have an understanding of the concept.

I borrow my approach to concept mastery from one of the greatest physicists of all time, Richard Feynman. I should probably mention some of the things that Feynman has accomplished: he's a Nobel Prize winner, was a part of the Manhattan Project to develop American nuclear weapons, and I could go on and on about his science accomplishments. He's also one of the most popular lecturers of all time. But back to concepts: his approach was to close the book, pick a topic, and on a blank sheet of paper write down everything he knows about it, and just explain it as if he were explaining it to someone who knows nothing about the subject. If you try this, at a certain point, it will become obvious that there are things that you don't understand. At that point, either do more reading, ask a question, or get help on that specific topic.

Figure 5.1 Sample concept map for mechanics

Albeit more time consuming, another approach to deepen your understanding is to construct a concept map. This approach is great for visual learners or creative people. Get started by taking the book or the lectures or both, and drawing a picture showing the main subjects. The main title of a chapter should be in the center of the page, and then you may have sections that break out of from center. Each section may also include several subsections which all contain small sketches of equations, pictures, or other visual cues that stimulate your creativity and memory.

An example of a concept map is shown in Figure 5.1. You should find creative ways to add formulas and sketches to make sure you understand the key concepts. This is really a great technique when you want to remember formulas, or if you want to remember how things fit together. And this is also a great way to create legal cheat sheets commonly referred to as "crib sheets." In physics courses or math courses you frequently get one note sheet of equations that you can make and bring to exams.

A lot of people make the mistake of trying to write as small as they can, trying to put as many homework problems as they can on that sheet. But the real value of crib sheets is the process of creating one: spending the time to organize the information in a way that makes sense to you. Once you have created your concept maps and crib sheets make sure you to file them away and come back to them as it's a great way to study for comprehensive exams.

Coming Back From a Horrible Start

Inevitably, no matter what you do, you're going to hit a setback in your studies. You're going to fail. You're going to be embarrassed. You're going to want to quit. And yet you have to push through. You must take a step back and assess—and of course this is true for life in general—what's going wrong. Is it your inability? Are your goals clear? Are you behind? Did you do your prework? Are you doing your homework? Are your notes structured effectively? Are you actually reading? Are you mastering the material? You really do have to think effectively to get to the root of the problem.

More often than not you just didn't prepare hard enough, and sometimes life gets in the way of you being focused. After you've identified what went wrong, the second step is to communicate that to people. You want to let the teacher know, "Hey, I clearly failed the exam, or I've missed some assignments, my goal is to pass this course, and in addition to passing this course I want to make sure that I learn the material." Because it's not all about the grades. People want to focus on the grade, and obviously passing with high marks is important, but actually learning is even more important.

The third step is changing your schedule. You're going to have to make some changes in your schedule to accommodate the extra work you're going to need to do to improve your grades. This is where it gets really tricky, because students in college oftentimes just give up and withdraw from the class, which costs thousands of dollars. So if you ever have the option of possibly passing the course, don't withdraw. But you do have to change your schedule. You have to either eliminate your social time if you've been spending too much time with your friends, or if you have a part-time job, you might need to take some time off or reduce your hours. If there's a sport, you might want to talk to your coach and see if they can help you find some extra time for the work that you're going to need to do.

Once you've communicated with your instructor and asked for some help, then it's time to review where you missed the mark. It's a waste of time to just go through the motions and get something wrong, but not go back and understand where you went wrong. Make it a point to redo every test and exam problem, to try to master it.

Then go back and work the system: get extra help and pay extra close attention. Make sure you're getting everything you can get out of the class, make sure that you're reading multiple books. If you can afford a tutor, get one. If extra help is available, go to the extra help. Don't seek out any extra resource without having doubled down on your individual effort.

"E"ngraining

Engraining in an academic context is essentially preparing for an exam. This is where you find out if you have what it takes to win. Over and over in life, we are faced with situations where we have no choice but to perform. You have to be able to perform on the basketball court, in the classroom, and beyond. Your ability to perform is really just a measure of your preparation.

First and foremost, under no circumstances do

you cheat. If you take the easy way out you cheat yourself, you don't learn anything. Secondly, is procrastinating to the point you end up cramming for a test. It's proven that when you cram you forget all of the information almost instantly, which is completely counterproductive to our goal of becoming educated. So my advice is to take your time and digest the material at a comfortable pace.

Many students don't thoroughly study for exams. They show up hoping to get the perfect blend of problems that they are familiar with. Don't be one of them. You have to put yourself in the frame of mind that you were so disciplined in your preparation, it doesn't matter what they're going to give you. This is really hard work. When it's time to perform, you have to understand the material so well that the information just spills out of you—that's such a great feeling.

When you get ready to start preparing for an exam, remove all distractions. You want to get rid of your cellphone, at least during your study sessions, close your email, and your social media. For the love of God, turn off the T.V. Some people study better with music; that hasn't ever really been a thing for me. Never set your study sessions for more than two hours max on a particular subject. In terms of scheduling, you want to give yourself at least five days before an exam. Anything more than a week tends to make a person have so much time that they are able to procrastinate.

It's also good to spend the majority of your time studying alone, say 70 percent, and up to 30 percent of your time working in a group—if possible. The group serves to reinforce what you already know: you'll often explain and ask questions on the material, all of which bolsters your understanding.

At one week before the exam, start committing your different concept maps to memory. Look at a chapter and all the problems associated with it and be able to work through them, so you've answered all your questions. All you have to do is remind yourself about minor details and stay fresh on the material.

Next, after you completed the concept mastery,

there's concept review. Here's where we go a level deeper and we're just making sure that we know all the formulas and we can explain these concepts to other students. In the couple days before the exam, start your semi-final review, in which you bring all of your concepts together. If there are five chapters, you've already reviewed all five, you've memorized what you need to memorize, and now on this third or fourth day of studying you're just skimming the material, just to make sure that it sticks very well. Perhaps do some "advanced drills" to ingrain the material The night before the exam, take a brief look of 30 or 40 minutes, just to make sure it stays fresh, and make sure you get a lot of sleep.

On the day of the exam, I always like to arrive early, making sure that I'm thinking positively, not worried. When you're prepared, you're anticipating, you're looking forward to the exam. Not only to get it over with, but also to show off and flex your muscles a little bit. I find that exams only make you really nervous if you didn't put the hours in and prepare like you should have.

As you take the test, read through all the

questions first to let your subconscious mind start to think about them. Then start with the easiest questions as a warmup. Move into the more difficult questions, and make sure that you monitor your time and leave at least five minutes to review the entire test so that you can try to catch any mistakes before you turn it in. Personally, I never turn a test in early. I sit there and I work and recheck and recheck until there's no more time left.

Inevitably you're going to encounter exam questions that you don't understand, or you're not sure of your answer. Maybe the algebra or the math isn't quite working out, or you're not even sure where to start. It can happen. In those cases, what I always like to do is (time permitting), I write everything out that I know about the situation in paragraph form. You might even want to write a question about what you don't know, what assumptions one needs to make, etc. This approach is good, because it gives you a starting point for conversations with the instructor after you get the exam back.

Chapter 7: What Did We Learn?

If you've read this book, you've seen that we tried to accomplish many things in a really short text, so I didn't give you a lot of fluff in here. I tried to point out some of the key principles that make students and student-athletes more successful. This book was written with student-athletes in mind, but it really applies to anyone who seeks to live a successful life.

We began this book talking about character. At a really high level, your character is everything that you do, the sum of your actions. And so, character begins with your outlook, your thoughts, and the way you approach life in general. Do you approach it from a really narrow perspective, where you only think about or value a few things? Or do you approach life from a holistic perspective, where you place a considerable amount of value on learning, improving, and growing? I tried to show you that character and your outlook on life are two of the major keys to success.

Within the concept of character, I suggest that a major failure in many students is their lack of curiosity. People who are really curious generally do well when they're focused. I just want to say one last time that there are *so many* topics in this universe that a person could be interested in! When you see something that you don't understand, or that's completely foreign to you, don't worry. Slow down, take a second, and think about it. Google it, read a book, ask a question, but don't move forward in ignorance.

One of the other things that we talked about is personal accountability. There's a lot going on in this country with school systems in trouble, poor teachers, and poverty all to blame, but there's only one person who is going to save you—and that's you. I mentioned that a minimal effort won't cut it and just to be clear that's going to class and doing your homework. Unfortunately, many students do less than the minimum by spacing out during lectures and missing assignments. However, if you are attentive, get your work done, are curious about the things you don't know, and you're willing go the extra mile, YOU WILL SUCCEED.

I rounded out our discussion of character with leadership. You have to put yourself in a position to lead others, and that comes with a bit of vulnerability, but that allows you to get the most out of every situation. So if you're on a basketball team, make it your job to not only improve yourself but help your teammates. If you're in a class, make it your job, not only to get an "A," but to join a study group to teach someone who's having trouble. Become an activist in your community. If you see a problem, get involved. Whenever you join other people, try to prepare yourself to take a leadership role. By being a major contributor and eventually working up to a leadership role you'll always get the most out of any activity, and that's going to take you a very, very long way in your life.

We spent quite a bit of time talking about individuals who just don't get it, who don't see the big picture. In my opinion, that lackadaisical attitude comes from not having a life plan. The first step in having a life plan is to realize exactly what you want your life to be. Whenever I meet new students I like to ask them open-ended questions, like what professions they are interested in, what they like to do, and what things they want to accomplish in life. They rarely have clear answers because they don't spend time thinking about these questions and writing down the detailed plans. When you're 13 or 14, you start to have an understanding of the world and you can begin to answer some of these questions by considering the aspects of life that you want to avoid. Obviously, goals and plans will change but we must write them down and revisit them as often as possible.

We also talked about using things that you love to foster your learning. That means you're going to have to tackle some very tough courses in school. What I tried to do in this book is show that a lot of students hit physics and just completely freak out because they think it's so difficult and they're just not into it. But what if you could connect physics to something that you love, like basketball? Then the subject becomes a bit more accessible because you're starting with a sport you love and know well. Once you master the basics, a deeper understanding may even reveal how one might improve their game. I understand that it might not be physics for you, it might be music, art, or another science. I'm telling you that you must use what you already know, the things that you love (like basketball) to serve as metaphors, to be connectors to the things that you don't know and that are challenging for you (like physics). Because there are so many rewards in life for us if we master the ability to learn the subjects that most people struggle with.

Learning physics is a great place to start is because it stimulates your brain to create more grey matter. Grey matter is brain tissue that directly affects muscle control and sensory perception such as memory, emotions, speech, decision making, and self-control. The main point here is that as we push ourselves to master tough subjects like physics, math, music, or even chess, we tend to increase the capacity of our brains to accomplish mastery in seemingly unrelated areas, including sports. Perhaps one great example of this is former NFL player John Urschel, who was named on the Forbes "30 under 30" list of outstanding young scientists. His research in mathematics allowed him to win several academic awards while playing guard for the Baltimore Ravens. Urschel is currently pursuing his Ph.D at MIT.

Intro to Physics

In Chapter 2 we introduced you to some basic physics concepts. In this chapter, we gave you the background to talk about basketball from a physical perspective. At the very beginning of the chapter, we started out talking about the branch of physics called *mechanics*, and about particular forces. I know that many of you guys were familiar with things like the force of gravity, maybe tension and springs, and things along those lines, but what you probably weren't familiar with were things like the normal force, or how friction worked, or the fact that gravity actually exists between all massive objects. So we gave you a bit of an introduction to that and then showed you how that's related to weight, and how those things start to connect to the sport of basketball. We discussed that the goal of mechanics problems frequently boils down to finding the equation of motion, because once you know the equation of motion, you can specify the state of the system at all times in the future.

Then we started talking a bit about linear motion, and the key to understanding linear motion was *vectors*. We learned vectors have both magnitude and direction and that scalars are quantities that only have a magnitude (like mass, temperature, etc.). We discussed velocity, position, and acceleration in two dimensions, which set us up for exploring kinematics, the geometry of motion. With this background, we were able to look at our first couple of example problems. We used kinematics to determine the high point (apex) of a free throw and the location the ball would fall to the ground. Through this process we determined the position of the ball as a function of time. These kinds of computations fall into a realm of kinematics called projectile motion, fundamental for the analysis of anything flying through the air.

We then moved on to the topic of angular motion, which is similar to linear motion but employs a different coordinate system. Instead of the familiar X and Y, we specify coordinates in terms of r, which is the radius, and θ, which is the angle of rotation. Then we introduced torques that behave very similarly to forces—torque is a twisting force that tends to cause rotation.

We spent a little bit of time talking about the conservation of momentum. Momentum analysis comes into play when you have two objects that strike one another, because at the moment that the two objects strike there's a spike in the force. This spike in the forces makes the sum of forces no longer continuous, so we can't use Newton's Second Law to get an EOM anymore. Instead, in this situation, we consider the objects' momentum, defined as the objects' mass times its volume. To solve the system, we use the law of conservation of momentum, which tells us that the initial momentum of the entire system is the same as the final momentum of the entire system.

That's the same whether we're talking about two basketballs that hit one another or several cue balls on a pool table. The main thing here is that the momentum at the beginning is the same as the momentum at the end, so that means if we know the momentum at the beginning (in particular the initial orientation of the system), we can solve for the mass or the final velocity because we know the momentum at the end. You can solve a lot of homework and exam questions that way.

Conservation of energy analysis is very similar to conservation of momentum. In order to analyze the energy it's crucial to understand that kinetic energy is the energy of something moving. Further, potential energy is the energy that's stored in an object and is related to its position in space. The example we considered was a basketball on the floor having no potential or kinetic energy, compared to a basketball on the top of a scoreboard having a high potential energy. Potential energy is actually stored in the second basketball because it can drop to the floor and generate speed.

Conservation of energy says that the energy is neither created nor destroyed, so the energy at the beginning is the same as the energy at the end, and it allows us to solve problems in a way that's analogous to a conservation of momentum problem.

Physics of Basketball

In Chapter 3, we studied the physics of basketball. I started out showing you why beginners have trouble with layups. Basically we talked about an inertial reference frame, and how that relates to a person running in towards the basket and releasing the ball. We talked about the fact that the ball will move in the horizontal direction at the same rate at which the body does. Players are often unaware of the consequence of inertia and add too much force in the direction of the basketball, causing the ball to slam off the backboard or front of the rim. This phenomenon also affects many other shots in that are taken on the move. We talked about the "runner," where a person tries stop his momentum at an instant and why aiming toward the center of the basket is not a good idea. One must aim to the side of the rim which is opposite to your motion to increase your chance of hitting that shot.

We investigated spin, why shots with backspin are stickier, or even luckier. I thought that the best way to show you this phenomenon would be to analyze what happens when we do a bounce pass. We gave you a situation when we did a bounce pass with three different types of spin, with forward spin, backspin, and no spin. And then you got a chance to see how the friction force on the floor changes due to the spin that's placed on the ball, which causes there to be a change in the transition between translational and rotational modes, which causes the ball to react completely differently. When you bounce pass the ball with forward spin, the ball does gains speed, and bounces to a lower height. When you bounce pass the ball with backspin, you lose speed, and it rebounds at an even higher height. Sometimes you can finesse this to the point where you put enough backspin on it that the ball actually bounces backward at a higher height. And then the intermediate step is to bounce a ball with no spin, and that will often rebound at a very similar angle to the initial incidental angle of the pass.

One thing I just wanted to clarify overall is that I probably didn't teach you guys too many new things in this physics of basketball chapter, especially you guys who play on high school basketball teams or college teams, or something along those lines. You have intuitively learned these things through the hours and hours that you have spent playing, but you didn't know how to describe it with physics. How cool is it to get out on the court and understand that the spin of the ball is related to its angular momentum, which is related to the frictional force from its hit on the floor? After you take your physics class, and when you start to hear those terms in the real world, you'll actually be able to think about what formulas are connected with real-life applications, so you could start to describe them mathematically. My reason for teaching you guys physics was to serve as an analogy. I love physics, but that doesn't mean you have to love it as much as I do. But it's immensely valuable for you to connect your knowledge with the curiosity for something new, and to know that everything in this world is related. There's value in learning things that you don't even think you want to know.

Improvement Philosophy

In Chapters 4 and 5, we spent a long time talking about my philosophy for self-improvement. I gave you this acronym G.A.M.E., which is Goal, Assess, Master, and Engrain. We talked about making sure that your goals are SMART goals, a very popular system in the business world. That's another acronym for you, an acronym inside an acronym. That is, you want to make your goals Specific, you want to make them Measurable, so you can take data and track whether or not your goal has been met. You want to make them Actionable. That means you want to set a goal and then take action. And you want to make sure your goal is Results-oriented. So, you're following up with your measurable goals and you judge your success on the actual results that you get. And you want to keep it Time-bound. You want to set a time limit, review, reset the time limit, change the goals. This is a thing that happens over and over, and if you get into the process of doing this consistently, then you live a lifestyle that lives by a plan, you have a clear set of goals, and that is what gives you the energy to go the extra mile and be accountable.

After the "G" in G.A.M.E., we talked about your Assessment. Your assessment requires you to deconstruct your goal into smaller parts, so that you understand what's important. In basketball you might break your offense down into close shots, medium shots, 3-pointers, shooting off of the dribble, shooting off of the screen, and things like that. You can break it into a dozen little parts. So you break it into a few key skills, I suggested six. Figure out the six most important skills to master, and then you come up with a method for measuring how good you are in this particular area. In order to figure that out, you could research online, you can contact an expert, or you can consult a book. There are a million different ways for you to figure out what assessments you can use to gauge how good you are at these different skill levels. In this book I gave you a few different types.

The next piece of G.A.M.E. was Mastery. We said that once you complete the assessment and discover your weaknesses that you should choose one or two aspects to work on that are going to have the most impact. If you want to improve in a particular area, you need to get clarity on the fundamental concept. In school this is very easy to do because you get direct feedback from teachers. In sports you might have to use some other resources to figure that out. But once you know what you're doing incorrectly, the logical next step is to repeat isolated drills that strengthen the skill. The work you do in mastery is a critical part of your pledge to go the extra mile, which ensures you are improving in the most efficient manner.

Last comes the "E," Engraining. You've worked on a skill in isolation, but does that mean you can go out and perform? How many times have you heard people say, "I studied for a test and I still failed," or, "I've been practicing but I'm not getting shots during the game?" That's because to truly engrain a skill, you have to practice *outside of your comfort zone*. You have to be able to put pressure on yourself. We talked about how to do that and I gave you some ideas on how to practice beyond your current skill level and how to replicate game-like conditions. For sports, this can mean scrimmaging against people who are far better than you. Try to practice at a higher level than you're going to be tested and watch yourself start to perform better than you ever have.

G.A.M.E. for Academics

In Chapter 6, we applied my improvement philosophy to the all-important area of academics. You know that the first step in G.A.M.E. is to set your Goals. To up your school G.A.M.E., think about whatever it is that you want to achieve in life and remind yourself that academic excellence will only help you to achieve this goal. Look into the future and see that it is extremely important for you to take advantage of the time given to you, as a student. Very seldom (or never again) will you have as much time to focus on your academics as you do now. That's why this is such an important time.

The second step in G.A.M.E. is Assessment of yourself. This is the easiest part, because school is all about assessments! Between your homework, projects and tests, you're being assessed all the time. Once your assignments are graded, you know where it is that you have to improve. I also gave you a few techniques ("drills") to improve your academic performance. Let's review a couple of those.

We talked about doing the prework. This is such a crucial step, and gives you SUCH an edge over the competition. All it takes is grabbing your textbook, getting the syllabus for the course *ahead of time*, and skimming. You're skimming through the chapters of the required textbook, though you could take it to the next level and skim over multiple books and create mind maps before the course even starts. But at a basic level you're just skimming everything so you're not hearing it for the first time. It takes dedication to do this during vacation but it will help you so much during the semester.

When it comes to homework, you want to understand how much the homework affects your grade, and you don't really want to spend more time doing homework than you have to, even if you score a bit lower on it. Of course, you want your homework completed, you want to give it your best shot, but you also need to spend time focusing on building up the areas that you're really weak in. We went into a lot of details, so they don't need revisiting here. But make sure that you're using other resources, asking for help, working with study groups, and things like that to get your homework done as fast as possible.

Also we talked about how to best use your time in class. Make sure that you never go multiple days without knowing what's going on in lecture. That is a cardinal sin. You *always* want to understand the main points of lecture. If you've been listening to a lecture for 20 minutes and you don't know what's going on, just raise your hand and say "Hey I'm sort of lost, what is the main point, what is the main step?" And before you leave class, make sure you ask the instructor about the variations of the problem, the different ways that this example can be worked out, and how you should decide how to proceed when solving the problem on your own.

It's so important that you ask those questions.

We talked about effective reading. So many students waste time reading the textbook word by word just trying to understand every little bit they can. I suggested a couple of different approaches. One approach is to get multiple books. Skim through three books on the same subject, use your class notes, and try to get as much as you can from there. Alternatively, you could employ the power reading method wherein as you read each paragraph you circle the most important words. In most books 60% of the words are phrases that don't add any value, so circle all the words that add value and then make sure you understand the main point. Then create your concept maps using those power words.

Next component of G.A.M.E. is Mastery. Mastery is really where we get in. We're not talking about getting an assignment completed, we're talking about coming up with your own specific drills that will help you improve. This can be difficult if you don't know what you're doing, but it's easy, if you're working on a chapter and you're not doing well on a particular section, to go back and do extra problems in this section.

Another alternative is to use learning websites like Khan Academy. They allow you to take an assessment. With this assessment you can go back and choose problems that will allow you to work on just one area. This is so important, because what if you missed something that you were supposed to have learned two years ago? Let's talk about algebra. You're solving for equations, you're in a trigonometry class, and you're struggling. You barely got the homework done, but you still don't quite understand what's going on. It's because your algebra is weak. So mastery is the portion where we go back and we understand how to solve for X, no matter what type of equation it is, just break it down and you focus on that.

Let's also not forget about the "E" in G.A.M.E., Engraining. Once you start to get it, then we do things like work on more difficult problems, and then you work without any help. That's closing the book, it's creating your concept maps, it's testing yourself with difficult problems under time constraints. When you do all of these things, you take yourself to a higher level of competence.

In the chapter we talked about a few disasters that are always going to happen, and I just wanted to give you some tips on how to manage these. One thing we covered was formula anxiety. That's when you open up a page (or test!) and you get overwhelmed because there's so much going on. When this happens, you've got to slow down, make sure you understand what all the symbols and all the variables mean, look at it for a while, take a break, and come back.

We also talked about getting sleepy when you don't have the time to sleep. You put in all this time and effort and sometimes you're reading through something and you just can't keep your eyes open. What I want you to do in those circumstances is take a break and go do something you love to do, but set a timer. For 5-10 minutes go bounce a ball, shoot a few jump shots, take a shower, what have you, and then come back. If you're getting sleepy because you're stuck, hopefully you haven't waited until the last minute and you can always go get help or work with a group.

The last several things are just about being prepared. Get a good night's rest, eat well, show up early, set your mind to the point where the only thing you're seeing is winning and no defeat, have a positive attitude, and go out and do your best.

Bringing It All Together

Now I want you to bring it all together. Cardinal rule number one is to make sure you have one master schedule with everything on it. I want you to have classes on it, "your" practice on it, "their" practice on it, and you should remember what I mean by those terms. Cardinal rule number two is that you don't want to waste time. There's so much extra time throughout the school day that you don't want to waste. There is prep period, study hall, lunch sessions, time in your classes when the substitute teacher is there or you're given time to complete your work. Use your time wisely and do not waste a moment. If you're going to accomplish those goals you set out in Chapter 1, you don't have the extra time.

For your brain and body to function at their peak, you need adequate sleep, no less than six and up to eight hours. Make sure you're doing everything you can to get as close to that eight hours as possible—more if you're training like crazy and your body tells you that you need it. The key to getting a good night's sleep is to go to bed on time!

I can't stress this enough: you want to do your homework as efficiently as possible. Do as much as you can, if you get to a failure point after you've actually tried the problem, go get help. Go to the resource center, to other students, to your teachers, ask questions, get it done, and move on. You have content to master and exams to ace. If you don't need help, be a leader and help someone else.

A lot of students miss a key point. You think that you get the summers off, but you don't. We should spend the summer preparing for the next school year. You should at least skim through the syllabus and the required material in every class that you're going to take in the next year. If you want to really go above and beyond, create concept maps for each course. You could do this in as little as two hours per day. The summer is also the time where you might take an extra class to study for a standardized test like the ACT or SAT. You may even want take extra courses online or do your own side projects. Whatever you do, don't become stagnant and lose momentum.

On the athletics side, the summer is the time when you can join some other leagues and gain access to a higher level of competition. A lot of students play on travelling teams and really take their games to the next level. Just remember, other than a short vacation here and there, you have no days off!

I know what I'm saying in this book is not easy, but the concepts are simple. If you complete the steps and follow through, you will be successful in any goal that you want to achieve. I'm looking forward to seeing you do great things.

Now go get 'em!